Silje Berger

Evergreen broad-leaved woody species - indicators of climate change

AF001280

Silje Berger

Evergreen broad-leaved woody species - indicators of climate change

Südwestdeutscher Verlag für Hochschulschriften

Impressum/Imprint (nur für Deutschland/ only for Germany)
Bibliografische Information der Deutschen Nationalbibliothek: Die Deutsche Nationalbibliothek verzeichnet diese Publikation in der Deutschen Nationalbibliografie; detaillierte bibliografische Daten sind im Internet über http://dnb.d-nb.de abrufbar.

Alle in diesem Buch genannten Marken und Produktnamen unterliegen warenzeichen-, marken- oder patentrechtlichem Schutz bzw. sind Warenzeichen oder eingetragene Warenzeichen der jeweiligen Inhaber. Die Wiedergabe von Marken, Produktnamen, Gebrauchsnamen, Handelsnamen, Warenbezeichnungen u.s.w. in diesem Werk berechtigt auch ohne besondere Kennzeichnung nicht zu der Annahme, dass solche Namen im Sinne der Warenzeichen- und Markenschutzgesetzgebung als frei zu betrachten wären und daher von jedermann benutzt werden dürften.

Verlag: Südwestdeutscher Verlag für Hochschulschriften GmbH & Co. KG
Dudweiler Landstr. 99, 66123 Saarbrücken, Deutschland
Telefon +49 681 37 20 271-1, Telefax +49 681 37 20 271-0
Email: info@svh-verlag.de
Zugl.: Hannover, Gottfried Wilhelm Leibniz Universität Hannover, Diss., 2007

Herstellung in Deutschland:
Schaltungsdienst Lange o.H.G., Berlin
Books on Demand GmbH, Norderstedt
Reha GmbH, Saarbrücken
Amazon Distribution GmbH, Leipzig
ISBN: 978-3-8381-0026-5

Imprint (only for USA, GB)
Bibliographic information published by the Deutsche Nationalbibliothek: The Deutsche Nationalbibliothek lists this publication in the Deutsche Nationalbibliografie; detailed bibliographic data are available in the Internet at http://dnb.d-nb.de.

Any brand names and product names mentioned in this book are subject to trademark, brand or patent protection and are trademarks or registered trademarks of their respective holders. The use of brand names, product names, common names, trade names, product descriptions etc. even without a particular marking in this works is in no way to be construed to mean that such names may be regarded as unrestricted in respect of trademark and brand protection legislation and could thus be used by anyone.

Publisher: Südwestdeutscher Verlag für Hochschulschriften GmbH & Co. KG
Dudweiler Landstr. 99, 66123 Saarbrücken, Germany
Phone +49 681 37 20 271-1, Fax +49 681 37 20 271-0
Email: info@svh-verlag.de

Printed in the U.S.A.
Printed in the U.K. by (see last page)
ISBN: 978-3-8381-0026-5

Copyright © 2011 by the author and Südwestdeutscher Verlag für Hochschulschriften GmbH & Co. KG and licensors
All rights reserved. Saarbrücken 2011

Table of contents

Abstract 3
Zusammenfassung 5
1 Introduction 7
2 An ecological "footprint" of climate change 11
   2.1   Summary 11
   2.2   Introduction 11
   2.3   Material and methods 12
   2.4   Results 13
   2.5   Discussion 17
   2.6   Acknowledgements 18
   2.7   References 19
3 Distribution of evergreen broad-leaved woody species in Insubria in relation to bedrock and precipitation 23
   3.1   Abstract 23
   3.2   Introduction 23
   3.3   Study area 25
   3.4   Methods 26
   3.5   Results 28
   3.6   Discussion 31
   3.7   Conclusions 35
   3.8   Acknowledgements 35
   3.9   Zusammenfassung 35
   3.10 References 36
4 Palms tracking climate change 41
   4.1   Abstract 41
   4.2   Introduction 42
   4.3   Material and Methods 43
   4.4   Results 45
       4.4.1 Bioclimatic preferences in the native habitat 45
       4.4.2 Potential and realised distribution in the introduced range 47
       4.4.3 The new northernmost palm population from a global perspective 49
   4.5   Discussion 50
   4.6   Acknowledgements 52
   4.7   References 53
5 Bioclimatic limits and range shifts of cold-hardy evergreen broad-leaved species at their northern distributional limit in Europe 58
   5.1   Abstract 58
   5.2   Introduction 58
   5.3   Methods 59
       5.3.1 Selected native and introduced evergreen broad-leaved species 60
   5.4   Results 61
   5.5   Discussion 66
       5.5.1 European distribution and bioclimatic limits 66
       5.5.2 Impacts at the landscape level 68
   5.6   Acknowledgements 71

5.7 References 72
6 General discussion 78
6.1 Temperature increase – the responsible driver for range shifts? 79
6.1.1 *Ilex aquifolium* 79
6.1.2 *Trachycarpus fortunei* 80
6.1.3 *Prunus laurocerasus* 81
6.1.4 Limiting temperature parameters 82
6.1.5 Climate variables 83
6.2 Detection and verification of vegetation shifts with bioclimatic models 84
6.3 Native vs. introduced evergreen broad-leaved species 84
6.4 Impact of precipitation and bedrock 85
6.4.1 Impact of precipitation 85
6.4.2 Impact of bedrock 87
6.5 Evergreen broad-leaved species – indicators of climate change? 87
Conclusions 88
6.6 Impacts on the vegetation level (Outlook) 88
Observed changes 88
Possible future changes 89
6.7 References 90

Acknowledgements 96

Appendix 98
Appendix 1 100
Appendix 2 102
Appendix 3 106
Appendix 4 118
Appendix 5 120

Abstract

Evergreen broad-leaved species are at their northern boundary of distribution in Central Europe. On the global scale, low winter temperatures play an important role limiting the distribution of evergreen broad-leaved vegetation towards the poles. In recent years, a global warming trend has been observed; the increase in annual mean temperature in Europe is mainly due to rising winter temperatures. In this study it is documented that the ranges of indigenous as well as introduced evergreen broad-leaved species are expanding northward in Central and Northern Europe. Furthermore, limiting parameters of single species are identified and the recent range shifts of some of the cold-hardiest evergreen broad-leaved species, such as *Ilex aquifolium*, *Prunus laurocerasus* and *Trachycarpus fortunei*, are analysed, based on historical and updated field data, measured climate data and output from bioclimatic models.

Within the group of evergreen broad-leaved species addressed in this study, different biological mechanisms are demonstrated to play a role in limiting the single species' distribution at their northern range margins. However, the northern ranges of the investigated species are all limited by low winter temperatures in general, though at different threshold values and due to specific biological traits. Furthermore, precipitation and bedrock may also influence regional distribution patterns.

The northward range shifts of several evergreen broad-leaved species are in concert with gradually increasing winter temperatures. Furthermore, the single species range shifts documented in the field confirm the simulated changes in species' distribution, expected from a bioclimatic model for current, relatively moderate, climate change. It may be concluded that the same underlying process, i.e. climate change, is the responsible driver of the synchronous expansion of several evergreen broad-leaved species. At the landscape scale, this indicates a considerable change in the composition and structure of temperate deciduous forests in various parts of Europe.

Keywords: Bioindicators, exotic species, global warming, range limiting parameters, range shifts.

Zusammenfassung

Immergrüne Laubgehölze erreichen ihre nördliche Verbreitungsgrenze in Mitteleuropa. Weltweit limitieren tiefe Wintertemperaturen die polwärtige Verbreitung des immergrünen Laubwaldbioms. In den letzten Jahren hat sich ein globaler Erwärmungstrend manifestiert, der in Europa vorwiegend auf mildere Winter zurückzuführen ist. In der vorliegenden Arbeit wird dokumentiert, dass sich die Verbreitungsgrenzen sowohl indigener als auch eingeführter immergrüner Laubgehölze in Mittel- und Nordeuropa nordwärts verschieben. Limitierende Parameter der Verbreitung einzelner Arten, wie *Ilex aquifolium*, *Prunus laurocerasus* und *Trachycarpus fortunei*, werden identifiziert, und die aktuellen Arealverschiebungen werden unter Berücksichtigung historischer und aktueller Verbreitungsdaten, gemessener Klimadaten und Ergebnisse bioklimatischer Modelle analysiert.

Verschiedene biologische Mechanismen bestimmen die nördlichen Verbreitungsgrenzen der einzelnen, untersuchten immergrünen Laubgehölzarten. Allerdings wird die nördliche Verbreitung aller im Detail untersuchten Arten durch tiefe Wintertemperaturen limitiert; hierbei sind jedoch unterschiedliche Schwellenwerte entscheidend, die durch artspezifische Merkmale bedingt sind. Des Weiteren beeinflussen Niederschlag und Ausgangsgestein kleinräumigere Verbreitungsmuster der Arten.

Die nordwärts gerichteten Arealverschiebungen verschiedener immergrüner Laubgehölze verliefen synchron mit Zunahme der Wintertemperaturen. Die im Feld dokumentierten Arealveränderungen bestätigen Erwartungen, die sich mit Hilfe eines bioklimatischen Modells für die bisherige, relativ moderate, Erwärmung ableiten lassen. Aus den vorgelegten Ergebnissen kann geschlussfolgert werden, dass der Klimawandel als zugrunde liegender Prozess die synchrone Arealverschiebung verschiedener immergrünen Arten ermöglicht hat. Es deutet sich eine bedeutende Änderung sowohl in der Artenzusammensetzung, als auch in der Struktur sommergrüner Wälder in verschiedenen Teilen Europas an.

Keywords: Arealverschiebungen, Bioindikatoren, eingeführte Arten, globale Erwärmung, limitierende Parameter.

1 Introduction

Climate is one important factor determining the geographical distribution of plant species. In the last decades the earth experienced a global warming trend (IPCC 2007). The average surface temperature in Europe increased by 0.95°C in the course of the last century, mainly due to rising winter temperatures and with an increasing warming rate towards the end of century (EEA 2004, IPCC 2001). Global warming is expected to continue (IPCC 2007), and consequently substantial shifts in vegetation patterns are projected based on bioclimatic modelling (e.g. BAKKENES et al. 2002, THOMAS et al. 2004, THUILLER et al. 2005). In the last years an increasing number of studies substantiated already detectable consequences of climate change, so called "fingerprints of climate change" (e.g. WALTHER et al. 2001, WALTHER et al. 2002, PARMESAN & YOHE 2003, ROOT et al. 2003).

One of these "fingerprints" was detected in the southernmost part of Switzerland, where a number of evergreen broad-leaved species were able to spread into deciduous forests in the last decades (GIANONI et al. 1988, KLÖTZLI et al. 1996). This phenomenon, termed "laurophyllisation" (KLÖTZLI et al. 1996), was investigated by WALTHER (2000) in the area surrounding Lago Maggiore. The flora of central Europe is relatively poor in evergreen broad-leaved species compared to temperate regions of other continents. Repeated glaciation diminished the former diverse Tertiary flora (HÜBL 1988, MAI 1995). However, in the region of Lago Maggiore numerous evergreen broad-leaved species from all over the world have been (re-)introduced and cultivated (SCHRÖTER 1936, SCHMID 1956). Numerous fertile individuals in gardens and parks hence served as seed sources for spread and naturalisation. Although the cultivation of exotic species has a centuries old tradition in the area of Lago Maggiore, the spread of exotic evergreen broad-leaved species was first observed in the last decades. Milder winters and a distinct decrease in the number of frost days weakened the climatic constraints for evergreen broad-leaved species and allowed them to spread (WALTHER 2000).

Many of the naturalising species, especially the most cold-hardy ones, are frequently cultivated also in other parts of Europe. In this study, the potential of evergreen broad-leaved species to serve as bioclimatic indicators on the European scale is analysed. Low winter temperatures are limiting the global distribution of evergreen broad-leaved species towards the poles (e.g. BOX 1981, KLÖTZLI 1988, WOODWARD et al. 2004, POTT 2005). The climatic limits of some evergreen broad-leaved species have been applied to reconstruct climatic fluctuations of the past from the geological record (e.g. IVERSEN 1944). In this study the sensitivity to low winter temperatures is analysed and applied to track recent climate change in Central Europe, where most of the recent warming has taken place in winter.

The historical and current ranges of native evergreen broad-leaved woody species are considered as well as the expansion of exotic species and the temperature parameters limiting their ranges. To understand the species' European distribution patterns detailed information on further ecological requirements are necessary, and hence the influence of precipitation and geological bedrock is investigated. Finally, based on the obtained knowledge of the single species' range shifts and specific limiting parameters, the impact on the vegetation level is discussed, relating the results to palaeoecological evidence on the one hand, on the other hand discussing them in context of vegetation changes expected due to future climate change.

References

BAKKENES, M., ALKEMADE, J.R.M., IHLE, F., LEEMANS, R. & J.B. LATOUR (2002): Assessing effects of forecasted climate change on the diversity and distribution of European higher plants for 2050. *Glob. Chang. Biol.* **8**: 390-407.

BOX, E. O. (1981): Macroclimate and plant forms: An introduction to predictive modelling in phytogeography. Tasks for vegetation science 1. – Junk Publishers, The Hague.

EEA (2004): EUA Signale 2004. Aktuelle Informationen der Europäischen Umweltagentur zu ausgewählten Themen. – Amt für amtliche Veröffentlichungen der Europäischen Gemeinschaften, Luxemburg.

GIANONI, G., CARRARO, G. & F. KLÖTZLI (1988): Thermophile, an laurophyllen Pflanzenarten reiche Waldgesellschaften im hyperinsubrischen Seenbereich des Tessins. *Ber. Geobot. Inst. ETH, Stiftung Rübel, Zürich* **54**: 164-180.

HÜBL, E. (1988): Lorbeerwälder und Hartlaubwälder (Ostasien, Mediterraneis und Makronesien). *Düsseldorfer Geobot. Kolloq.* **5**: 3-26.

IPCC (2001): Climate Change 2001: The scientific basis. Contribution of working group I to the third assessment report of the intergovernmental panel on climate change. – Cambridge University Press, Cambridge.

IPCC (2007): Climate Change 2007: The Physical Science Basis. – IPCC, Genf. URL: http://www.ipcc.ch, 15.06.07.

IVERSEN, J. (1944): *Viscum, Hedera* and *Ilex* as climatic indicators. *Geol. Fören. Förhandl.* **66**: 463-483.

KLÖTZLI, F. (1988): On the global position of the evergreen broad-leaved (non-ombrophilous) forest in the subtropical and temperate zones. *Veröff. Geobot. Inst. ETH, Stiftung Rübel, Zürich* **98**: 169-196.

KLÖTZLI, F., WALTHER, G.-R., CARRARO, G. & A. GRUNDMANN (1996): Anlaufender Biomwandel in Insubrien. *Verh. Ges. Ökol.* **26**: 537-550.

MAI, D.H. (1995): Tertiäre Vegetationsgeschichte Europas. – Fischer, Jena.

PARMESAN, C. & G. YOHE (2003): A globally coherent fingerprint of climate change impacts across natural systems. *Nature* **421**: 37-42.

POTT, R. (2005): Allgemeine Geobotanik. – Springer, Berlin.

ROOT, T.L., PRICE, J.T., HALL, K.R., SCHNEIDER, S.H., ROSENZWEIG, C. & J.A. POUNDS (2003): Fingerprints of global warming on wild animals and plants. *Nature* **421**: 57-60.

THOMAS, C.D., CAMERON, A., GREEN, R.E., BAKKENES, M., BEAUMONT, L.J., COLLINGHAM, Y.C., ERASMUS, B.F.N., DE SIQUEIRA, M.F., GRAINGER, A., HANNAH, L., HUGHES, L., HUNTLEY, B., VAN JAARSVELD, A.S., MIDGLEY, G.F., MILES, L., ORTEGA-HUERTA, M.A., PETERSON, A.T., PHILLIPS, O.L. & S.E. WILLIAMS (2004): Extinction risk from climate change. *Nature* **427**: 145-148.

THUILLER, W., LAVOREL, S., ARAÚJO, M.B., SYKES, M.T. & I.C. PRENTICE (2005): Climate change threats to plant diversity in Europe. *Proc. Nat. Acad. Sci. USA* **102**: 8245-8250.

WALTHER, G.-R. (2000): Laurophyllisation in Switzerland. – Dissertation, ETH Zürich.

WALTHER, G.-R., BURGA, C.A. & P.J. EDWARDS (2001): "Fingerprints" of Climate Change – Adapted behaviour and shifting species ranges. – Kluwer Academic/Plenum Publishers, New York.

WALTHER, G.-R., POST, E., CONVEY, P., MENZEL, A., PARMESAN, C., BEEBEE, T.J.C., FROMENTIN, J.M., HOEGH-GULDBERG, O. & F. BAIRLEIN (2002): Ecological responses to recent climate change. *Nature* **416**: 389-395.

WOODWARD, F.I., LOMAS, M.R. & C.K. KELLY (2004): Global climate and the distribution of plant biomes. *Philos. Trans. R. Soc. Lond. B* **359**: 1465-1476.

2 An ecological "footprint" of climate change[1]

2.1 Summary

Recently, there has been increasing evidence of species' range shifts due to changes in climate. Whereas most of these shifts relate ground truth biogeographic data to a general warming trend in regional or global climate data, we here present a reanalysis of both biogeographic and bioclimatic data of equal spatio-temporal resolution, covering a time span of more than fifty years. Our results reveal a coherent and synchronous shift in both, species' distribution and climate. They show not only a shift in the northern margin of a species which is in concert with gradually increasing winter temperatures of the area, they also confirm the simulated species' distribution changes expected from a bioclimatic model under the recent relatively moderate climate change.

Key words: range shift, global warming, bioindicator, bioclimatic model, evergreen broad-leaved species, *Ilex aquifolium*

2.2 Introduction

Despite an increasing number of ecological "fingerprints" of climate change (WALTHER *et al.* 2001, PARMESAN & YOHE 2003, ROOT *et al.* 2003; see also HUGHES 2000, WALTHER *et al.* 2002), consensus on the ecological impacts of global warming is still considered to remain elusive (see JENSEN 2003). One reason for the lack of consensus may relate to the fact that case studies on species' range shifts (e.g. GRABHERR *et al.* 1994, PARMESAN *et al.* 1999, THOMAS & LENNON 1999, HILL *et al.* 2002, CROZIER 2003) often associate local changes in the distribution of species at small scales to large-scale climatic changes on the regional to global level. This is mainly due to the lack of historical biogeographic data on the local and regional distribution of a species coupled with concurrent climatic data on the same spatio-temporal resolution. One of the exceptions, and to our knowledge the only one, is provided by IVERSEN (1944), who closely linked the occurrence of some evergreen broad-leaved species to measurements from nearby climate stations. Thus it has been used as the classical example to illustrate a climatically limited species' northern distribution and continues to be regularly featured in standard textbooks of ecology in general (e.g. BEGON *et al.* 1996), and botany in particular (SITTE *et al.* 2002, LARCHER 2003).

---

[1] Published in Proceedings of the royal society of London series B.
Royal Society Publishing: http://publishing.royalsociety.org/
Full reference: WALTHER, G.-R., BERGER, S. & M. T. SYKES (2005): An ecological 'footprint' of climate change. *Proc. R. Soc. Lond. B* **272**: 1427-1432.

A particular feature of the IVERSEN (1944) study is that it provides a detailed synchronous record for the bioclimatic and biogeographic situation of climatically limited species before the recent rise in global average temperature. In recent decades, the regional climate of the Iversen study area has warmed, especially in the winter season (FOLLAND & KARL 2001). Given this warming trend, one would expect to see a truly climatically limited species to respond to the changes in climate in these areas.

In this paper we compare both detailed historic climatic and biogeographic records with updates of the same parameters at the same localities in the northern fringe area of the distribution of *Ilex aquifolium,* applying the same methodology as used in the original study (IVERSEN 1944). We also compare our field data to predictions from simulations of *Ilex* distributions using a purely climate driven bioclimatic model. Our aim is therefore to verify whether a potential shift in the local and regional distribution of a species is in synchrony with concurrent changes in climatic data on the same spatio-temporal resolution.

2.3 Material and methods

From among the subset of evergreen species described by IVERSEN (1944) we selected holly (*Ilex aquifolium*), the only true shrub and smaller tree species (PETERKEN & LLOYD 1967, CALLAUCH 1983). The northern margin of the *Ilex* distribution in Europe is often shown related to the 0°C-isoline (HOLMBOE 1913, LOESENER 1919, ENQUIST 1924, WALTER & STRAKA 1970, SITTE et al. 2002). Although the mean temperatures are not themselves considered to be physiologically effective, they correlate with absolute minimum, and therefore are used as a surrogate for the frequency of lethal extreme events (SITTE et al. 2002; WOODWARD et al. 2004, see also RUNGE 1950, FISCHER 1965, PETERKEN & LLOYD 1967, SAKAI 1982, PRENTICE et al. 1992, SYKES et al. 1996). The other evergreen species studied by IVERSEN (1944) either depend on appropriate host trees such as the epiphyte *Viscum album,* or, in the case of the climbing ivy (*Hedera helix*), have two separate and different growth forms. Ivy, when creeping on the ground, is protected from winter cold by snow and is thus, in this growth form a less reliable climate indicator than when it is climbing trees (see also ANDERGASSEN & BAUER 2002). Further, these two different growth forms have not been clearly separated in the original data (IVERSEN 1944).

Our biogeographic field data for holly were obtained from several unpublished recent records of *Ilex aquifolium* scattered through various local field surveys and monitoring programmes (see results and acknowledgements sections below). We verified these records in the field in the northern fringe area of the distribution of holly in northern Germany, Denmark, southern Norway and southern Sweden in 2003, and updated the same climatic parameters of the same local climate stations as used in the original study (IVERSEN 1944). As the study covers more than a half century, not all the climate stations can provide data to the present day. For the abandoned stations we used data from nearby stations or extrapolated surrogate records based upon overlapping periods (see Appendix 1).

The STASH bioclimatic model is a simple model which uses a minimum set of bioclimatic parameters (mean temperature of the coldest month, growing season warmth, drought index, and degree of chilling required before budburst) to describe a species range. The assumption is that these parameters represent responses to physiologically important mechanisms for plants, for example accumulated temperature during the growing season (growing degree days) is an index of energy suitable for completion of the plant's life cycle. Some of these parameters act directly as on-off switches for example if the minimum mean coldest month temperature (as a surrogate for absolute minimum) in a grid cell falls below the species limit, the species is excluded. Some parameters however also directly affect net assimilation and respiration and thus growth rate and this is reflected in the degree of establishment success within a grid cell (see SYKES et al. 1996 for further details). We used STASH in a data-model comparison for current and recent past occurrences of holly. We used 12 monthly values for each of monthly mean temperature, precipitation and percentage sunshine from the ATEAM European gridded 10' climate dataset (http://www.pik-potsdam.de/ateam/) based on the CRU dataset (NEW et al. 1999, 2000). The recent past was modelled as the 30 year normal for 1931-1960 period and the current as the mean of 1981-2000.

Statistical analyses of the simulated and realised distributions of holly in the area between 51°-70° N and 4°-25° E were based on Cohen's Kappa coefficient of agreement (COHEN 1960) providing an estimate for the overlap of spatial data (cf. also FIELDING & BELL 1997).

2.4 Results

At all the localities where Iversen reported the occurrence of *Ilex aquifolium*, we confirmed that the species is still present. In addition, we found new occurrences of holly (*Ilex aquifolium*) at locations, which were reported *Ilex*-free at the time of Iversen's investigation (cf. Fig. 2.1 A/C). The growth form and size, date of first notification, and at some places tree ring analyses allowed an estimation of the approximate age of these individuals (data not shown, but cf. Berger 2003). Based on this information we conclude that these occurrences of *Ilex* are in fact new, and that *Ilex* was not present at the time of Iversen's investigation. These new occurrences represent a geographic shift in the distribution of holly towards the north in Norway and north-east in Germany, Denmark (cf. also

BANUELOS *et al.* 2004) and especially southern Sweden, where *Ilex aquifolium* expanded into new areas along the southern Swedish coast (Fig. 2.1 D). In addition, the results of the field survey have been plotted in a thermal correlation diagram (Fig. 2.2), i.e. a coordinate system, where the ordinate represents the mean temperature for the warmest month (July/August, depending on the individual station), and the abscissa the mean temperature for the coldest month (January/February, same as for warmest month) (cf. IVERSEN 1944). In this thermal correlation diagram (Fig. 2.2) not only the status but also the position of the related climatic stations changed in accordance with the findings of the field survey. Whenever a station, previously designated holly-free, advanced to or even crossed the thermal limit curve in the thermal correlation diagram, field observations revealed that the station's surrounding area has been colonized by holly in the time since Iversen's investigation. Consequently, the change in the position of these stations is in synchrony with the change in their biogeographic status, indicated by different symbols in Fig. 2.2. Furthermore, while the position and status of several stations have changed, the position of the thermal limit curve remained stable compared to the outline given by IVERSEN (1944) (cf. Fig. 2.2). In the past, the 0°C January-isoline ran parallel with the northern margin of *Ilex*' distribution (Fig. 2.1 A), and with the recent shift of both, climate isoline and species' distribution change this relationship remains consistent (Fig. 2.1 C). Last but not least, the bioclimatic model based on purely bioclimatic response factors simulates a potential *Ilex* distribution of the recent past (based on 1931-60 climate data), that matches the realised area with a Kappa value [K] of 0.82, which is well in the range of "excellent" agreement (:= K > 0.75 according to FIELDING & BELL 1997, see also LANDIS & KOCH 1977). The model also predicts new areas to become colonised by holly under a warming scenario based on the 1981-2000 climate data (Fig. 2.1 D). In this case, the Kappa value for the simulated range shift and the observed change (the latter is the area delimited by the new occurrences (see triangles in Fig. 2.1 D) excluding the realised area of the recent past (see Fig. 2.1 A)) is 0.50, which, though lower, is in the range of a "moderate" Kappa agreement (LANDIS & KOCH 1977).

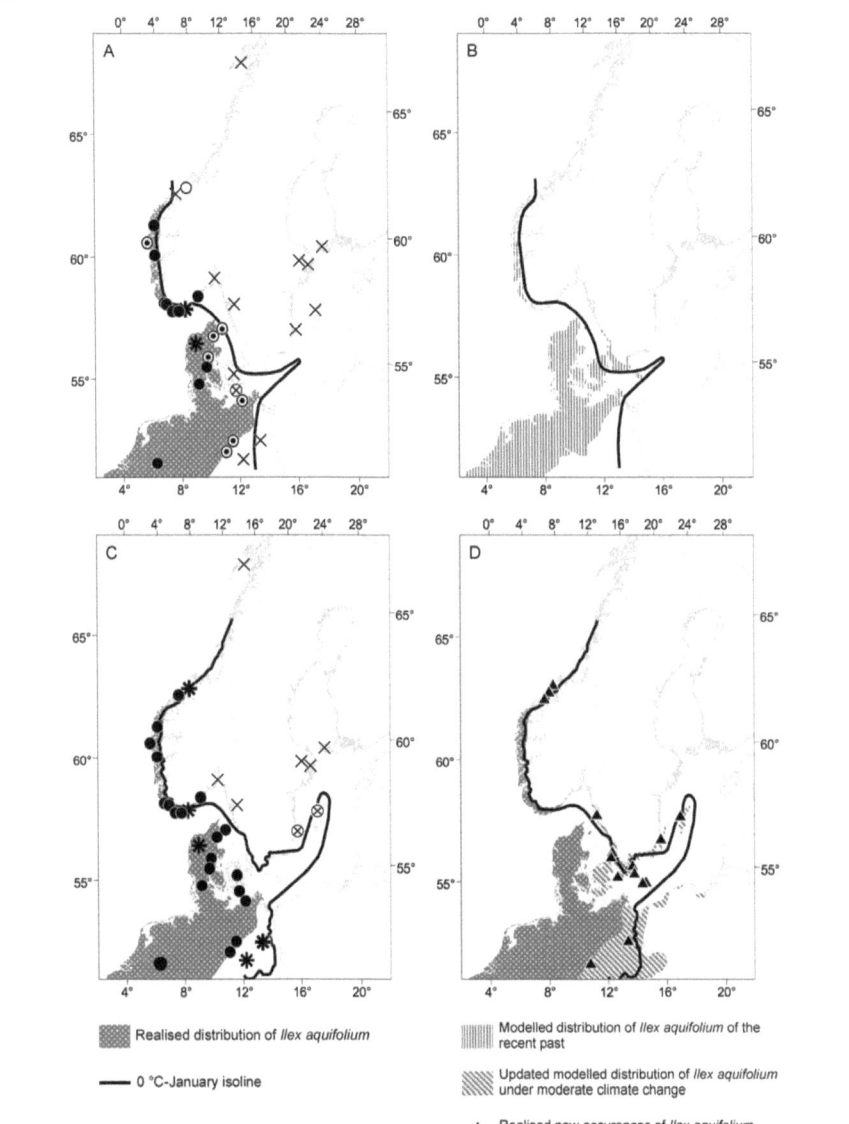

Fig. 2.1: Distribution of *Ilex aquifolium* and 0 °C-January isoline at different times.
A) Former range of *Ilex aquifolium* [dark grey shading] based on ENQUIST (1924) and MEUSEL *et al.* (1965), isoline based upon WALTER & STRAKA (1970), symbols based upon IVERSEN (1944), for symbol legends see Fig. 2.2. B) Modelled range of *Ilex aquifolium* of the recent past (1931-60) [vertical shading], isoline as in Fig. 2.1A. C) Former range of *Ilex aquifolium* [dark grey shading] as in Fig. 2.1A, isoline updated for 1981-2000 based on MITCHELL *et al.* (2004), symbols updated, for symbol legends see Fig. 2.2. D) Former range of *Ilex aquifolium* [dark grey shading] complemented with the simulated species' distribution under a moderate climate change based on 1981-2000 climate data [diagonal shading], isoline as in Fig 2.1C, triangles represent locations with realised new occurrences of *Ilex aquifolium*.

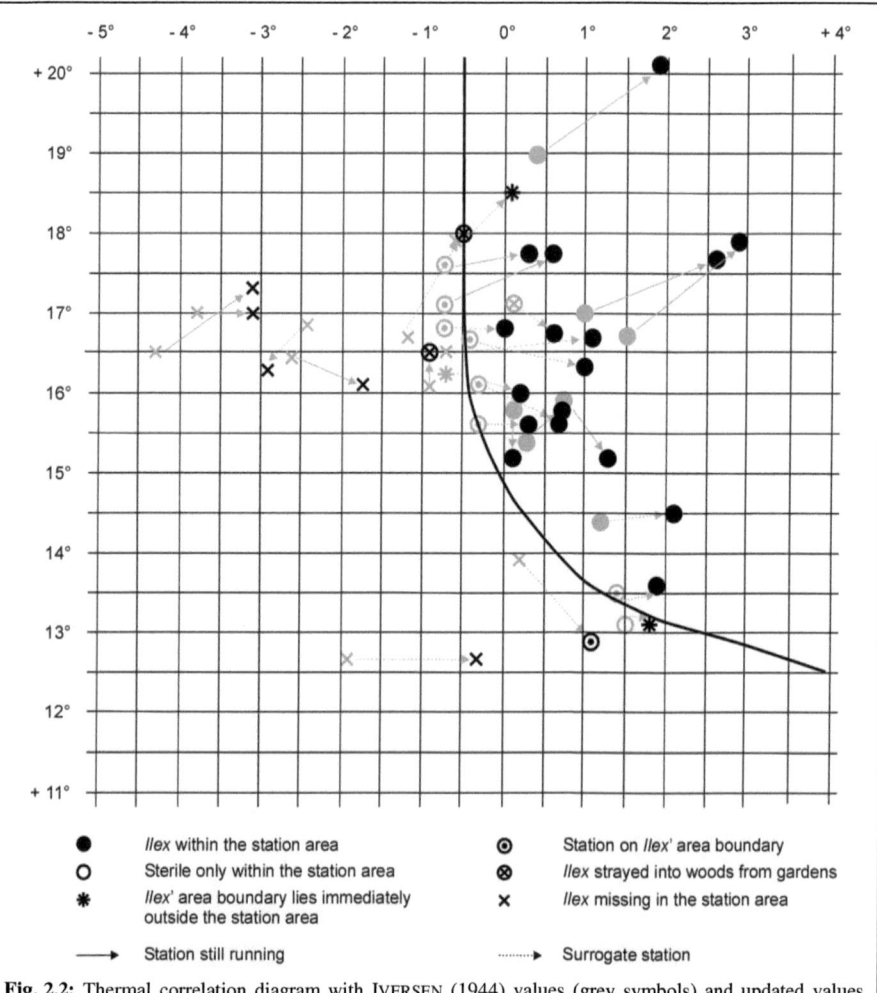

Fig. 2.2: Thermal correlation diagram with IVERSEN (1944) values (grey symbols) and updated values (black symbols); the arrows link the values of corresponding stations. For details see text and appendix.

2.5 Discussion

In the original work, the climatic stations were critically selected in order to ensure they sufficiently represented the surrounding areas (defined by IVERSEN (1944) as the area within a circle of 20 km and within a vertical distance of 40 metres). Because of the long period of this study (c. 50 years) it was not possible to update all the climate data from all the original stations. However, due to the large geographic distances between the climatic stations used in this study some minor variation arising from the use of nearby surrogate climate stations for abandoned original stations is relatively insignificant, especially when considering the potential differences in proportion to changes in microhabitats within the 20 km circle considered representative for the thermosphere of the plants (see IVERSEN 1944). Furthermore, the data used in the thermal correlation diagram revealed no systematic discrepancy in terms of magnitude or direction of change in status or position between the subset of original and surrogate stations (see Appendix 1). Therefore, although not all the original climate stations survived the time span covered in this study, the data set is considered sufficiently reliable for the purpose of the study. Also, the observed change especially in winter temperature accords well with the latest IPCC findings reporting a 0.6–1.0°C/decade warming experienced in the period 1976–2000 in southern Scandinavia (FOLLAND & KARL 2001).

Although we could have based our re-survey on records from several local vegetation monitoring programs, it was necessary to verify all notifications of (potential) *Ilex*-occurrences in the field, because for example in some places, we identified *Mahonia aquifolium* instead of the expected *Ilex aquifolium*. All the localities with previous occurrences of *Ilex* reported by IVERSEN (1944) were confirmed. In addition, we also found new areas with holly, that were reported *Ilex*-free at the time of Iversen's investigation. In some cases, new individuals were considered to be escapees from planted garden individuals. However, in that regard, Iversen in his study also included a category named *"Ilex* strayed into woods from gardens" (cf. IVERSEN 1944, p. 471). Such subspontaneous occurrences are not in opposition to our approach, rather they are in agreement with the methodology and findings of the original study. Such events provide opportunities for *Ilex* to keep pace with the rate of climate change by reducing the time-lag that may be due to, for example, dispersal limitations (SVENNING & SKOV 2004, cf. also POTT 1990, LEEMANS 1996) and/or interrupted migration routes (SKOV & SVENNING 2004) thus allowing a species the chance to occupy its new potential range immediately.

The observed north- and north-eastward range expansion tracks the increased warming measured at local climate stations. Whenever a station, which previously was reported *Ilex*-free, advanced to or crossed the thermal limit curve in the thermal correlation diagram, a new occurrence of holly was found in the field in the surrounding area. Furthermore, both the historic and the present northern margin of *Ilex* distribution in Europe remains related to the 0°C-isoline (Fig. 2.1 A/C). STASH model output showed very clearly that in the past, *Ilex* has filled a great portion of its potential range (Fig. 2.1 B). Furthermore, the new occurrences of *Ilex aquifolium* overlap with the potential range of this species under the recent moderate climate change predicted by the model using climate data of the last two decades only (see Fig. 2.1 D, cf. also BANUELOS et al. 2004 for Denmark). However, the lower Kappa value resulting from the comparison of the expected and the realised shift in distribution of the last two decades suggests that *Ilex*, probably for chorological reasons, has not yet fully occupied its climatically determined potential new range predicted by the model (cf. also SVENNING & SKOV 2004).

Northern range limitation by climatic parameters has not only been reported from plant species (e.g. WOODWARD 1987, GRAVES & REAVEY 1996, WOODWARD et al. 2004; see also WALTHER 2004), it is also of importance for other taxa such as insects (e.g. HILL et al. 1999, ADDO-BEDIAKO et al. 2000, THOMAS et al. 2001), birds (e.g. THOMAS & LENNON 1999, FORSMAN & MONKKONNEN 2003, BROMMER 2004) and mammals (e.g. HUMPHRIES et al. 2002), and thus, an important ecological feature in temperate regions. The present resurvey of the distribution of a climatically limited species and the reanalysis of closely related local climatic measurements revealed a coherent shift in both, species' distribution and climate with the same spatio-temporal resolution. This attributes high confidence to the conclusion, that the changing climate is the responsible driver for the observed northward range expansion. Consequently, this reported species' shift is more than just an ecological "fingerprint", it is an ecological "footprint" of recent climate change.

2.6 Acknowledgements

Floristic data were kindly provided by T. Tyler (Projekt Skånes Flora, Sweden), L. Jonsell (Projekt Upplands Flora, Sweden), S. Svensson (Municipality of Gotland, Sweden), C.-A. Hæggström (Department of Ecology, University of Helsinki, Finland), G. Weimarck (Göteborg, Sweden), P. H. Salvesen (The Norwegian Arboretum, University of Bergen, Norway), A. Skogen (University of Bergen, Norway), J. Kollmann & J. M. Bañuelos (Department of Ecology, Royal Veterinary and Agricultural University of Denmark, Copenhagen, Denmark), J. Lawesson (Institute of Biological Sciences, University of Aarhus, Denmark), G. Stachowiak (Salzwedel, Germany),

D. Frank (Halle, Germany), R. May (German Federal Agency for Nature Conservation, Bonn, Germany), W. Härdtle (Institute of Ecology and Environmental Chemistry, University of Lüneburg, Germany), M. Diekmann (Vegetation Ecology and Conservation Biology, University of Bremen, Germany); and climatic data by K. Boqvist (Swedish Meteorological and Hydrological Institute, Norrköping, Sweden), W. Ernst (Deutscher Wetterdienst, Offenbach, Germany), L. Ågren (Ecological Research Station of Uppsala University, Färjestaden, Sweden), S. Kristiansen (Meteorological Institute, Oslo, Norway), K. M. Erhardi (Danish Meteorological Institute, Copenhagen, Denmark) T. Mitchell (Tyndall Centre for Climate Research, Norwich, United Kingdom), T. Carter & S. Fronzek (Finnish Environment Institute, Helsinki, Finland). We also thank G. Marion and S. Bierman (BIOSS, Edinburgh, United Kingdom) for statistical advice and two anonymous referees for helpful comments on earlier versions of the manuscript. Funding by the following agencies is kindly acknowledged: German Research Foundation (Project WA 1523/5-1), and the EC (within the FP 6 Integrated Project "ALARM"; GOCE-CT-2003-506675).

2.7 References

ADDO-BEDIAKO, A., CHOWN, S.L. & K.J. GASTON (2000): Thermal tolerance, climatic variability and latitude. *Proc. R. Soc. Lond.* B **267**: 739-745.

ANDERGASSEN, S. & H. BAUER (2002): Frost hardiness in the juvenile and adult life phase of Ivy (*Hedera helix* L.). *Plant Ecol.* **161**: 207-213.

BANUELOS, M.J., KOLLMANN, J., HARTVIG, P. & M. QUEVEDO (2004): Modelling the distribution of *Ilex aquifolium* at the north-eastern edge of its geographical range. *Nord. J. Bot.* **23(1)**: 129-142.

BEGON, M., HARPER, J.L. & C.R. TOWNSEND (1996): Ecology – Individuals, populations and communities. 3rd ed. – Blackwell, Oxford.

BERGER, S. (2003): *Ilex aquifolium* – Bioindikator für Klimaveränderung? – Diplomarbeit, Institut für Geobotanik, Universität Hannover.

BROMMER, J.E. (2004): The range margins of northern birds shift polewards. *Ann. Zool. Fenn.* **41**: 391-397.

CALLAUCH, R. (1983): Untersuchungen zur Biologie und Vergesellschaftung der Stechpalme *(Ilex aquifolium).* – Dissertation, Universität des Landes Hessen, Kassel.

COHEN, J. (1960): A coefficient of agreement for nominal scales. *Educ. Psychol. Meas.* **20**: 37-46.

CROZIER, L. (2003): Winter warming facilitates range expansion: cold tolerance of the butterfly *Atalopedes campestris. Oecologia* **135**: 648-656.

ENQUIST, F. (1924): Sambandet mellan klimat och växtgränser. *Geol. Fören. Förhandl.* **46**: 202-213.

FIELDING, A.H. & J.F. BELL (1997): A review of methods for the assessment of prediction errors in conservation presence/absence models. *Environ. Conserv.* **24**: 38-49.

FISCHER, W. (1965): Über Wassergehalt und Standortsverhältnisse bei einigen wintergrünen atlantischen Pflanzenarten an der Ostgrenze ihrer Verbreitung in NW-Brandenburg. – Dissertation, Dtsch. Akad. Landwirtschaftswiss., Berlin.

FOLLAND, C.K. & T.R. KARL (2001): Observed Climate Variability and Change. In: HOUGHTON, J.T., DING, Y., GRIGGS, D.J., NOGUER, M., VAN DER LINDEN, P.J., DAI, X., MASKELL, K. & C.A. JOHNSON (eds.): Climate Change 2001: The Scientific Basis. pp. 99-181. – Cambridge University Press, Cambridge.

FORSMAN, J.T. & M. MONKKONEN (2003): The role of climate in limiting European resident bird populations *J. Biogeogr.* **30**: 55-70.

GRABHERR, G., GOTTFRIED, M. & H. PAULI (1994): Climate effects on mountain plants. *Nature* **369**: 448.

GRAVES, J. & D. REAVEY (1996): Global Environmental Change – Plants, Animals and Communities. – Longman, Essex.

HILL, J.K., THOMAS, C.D. & B. HUNTLEY (1999): Climate and habitat availability determine twentieth century changes in a butterfly's range margins. *Proc. R. Soc. Lond.* B **266**: 1197-1206.

HILL, J.K., THOMAS, C.D., FOX, R., TELFER, M.G., WILLIS, S.G., ASHER, J. & B. HUNTLEY (2002): Responses of butterflies to twentieth century climate warming: implications for future ranges. *Proc. R. Soc. Lond.* B **269**: 2163-2171.

HOLMBOE, J. (1913): Kristtornen i Norge. *Bergens Museums Aarbok* **7**: 1-91.

HUGHES, L. (2000): Biological consequences of global warming: is the signal already apparent? *Trends Ecol. Evol.* **15**: 56-61.

HUMPHRIES, M.M., THOMAS, D.W. & J.R. SPEAKMAN (2002): Climate-mediated energetic constraints on the distribution of hibernating mammals. *Nature* **418**: 313-316.

IVERSEN, J. (1944): *Viscum, Hedera* and *Ilex* as climatic indicators. A contribution to the study of past-glacial temperature climate. *Geol. Fören. Förhandl.* **66**: 463-483.

JENSEN, M.N. (2003): Consensus on ecological impacts remains elusive. *Science* **299**: 38.

LANDIS, J.R. & G.C. KOCH (1977): The measurement of observer agreement for categorical data. *Biometrics* **33**: 159-174.

LARCHER, W. (2003): Physiological Plant Ecology. 4th ed. – Springer, Berlin.

LEEMANS, R. (1996): Biodiversity and global change. In: K.J. Gaston (ed.): Biodiversity – A Biology of Numbers and Difference. pp. 367-387 – Blackwell, Oxford.

LOESENER, T. (1919): Über die Aquifoliaceen, besonders über *Ilex. Mitt. Dtsch. Dendrol. Ges.* **28**: 1-66.

MEUSEL, H., JÄGER, E. & E. WEINERT (1965): Vergleichende Chorologie der zentraleuropäischen Flora. – Fischer, Jena.

MITCHELL, T.D., CARTER, T.R., JONES, P.D., HULME, M. & M. NEW (2004): A comprehensive set of high-resolution grids of monthly climate for Europe and the globe: the observed record (1901-2000) and 16 scenarios (2001-2100). *Tyndall Centre Working Papers* **55**: July 2004, available at: http://www.tyndall.ac.uk/publications/working_papers/wp55.pdf

NEW, M., HULME, M. & P.D. JONES (1999): Representing twentieth century space-time variability. Part I. Development of a 1961-90 mean monthly terrestrial climatology. *J.Clim.* **12**: 829-856.

NEW, M., HULME, M. & P.D. JONES (2000): Representing twentieth century space-time variability. Part 2. Development of 1901-96 monthly grids of terrestrial surface climate. *J. Clim.* **13**: 2217-2238.

PARMESAN, C., RYRHOLM, N., STEFANESCU C., HILL, J.K., THOMAS, C.D., DESCIMON, H., HUNTLEY, B., KAILA, L., KULLBERG, J., TAMMARU, T., TENNET, W.J., THOMAS J.A. & M. WARREN (1999): Poleward shifts in geographical ranges of butterfly species associated with regional warming. *Nature* **399**: 579-583.

PARMESAN, C. & G. YOHE (2003): A globally coherent fingerprint of climate change impacts across natural systems. *Nature* **421**: 37-42.

PETERKEN, G.F. & P.S. LLOYD (1967): Biological flora of the British Isles. *J. Ecol.* **55**: 841-858.

POTT, R. (1990): Die nacheiszeitliche Ausbreitung und heutige pflanzensoziologische Stellung von *Ilex aquifolium* L. *Tuexenia* **10**: 497-512.

PRENTICE, I.C., CRAMER, W., HARRISON, S., LEEMANS, R., MONSERUD, R.A. & A.M. SOLOMON (1992): A global biome model based on plant physiology and dominance, soil properties and climate. *J. Biogeogr.* **19**: 117-134.

ROOT, T.L., PRICE, J.T., HALL, K.R., SCHNEIDER, S.H., ROSENZWEIG, C. & J.A. POUNDS (2003): Fingerprints of global warming on wild animals and plants. *Nature* **421**: 57-60.

RUNGE, F. (1950): Die Standorte der Hülse (*Ilex aquifolium* L.) in der Umgebung des Naturschutzgebietes "Heiliges Meer" bei Hopsten (Westf.). *Natur und Heimat* **10**: 65-77.

SAKAI, A. (1982): Freezing Resistance of Ornamental Trees and Shrubs. *J. Am. Soc. Hortic. Sci.* **107**: 572-581.

SITTE, P., WEILER, E.W., KADEREIT, J.W., BRESINSKY, A. & C. KÖRNER (2002): Strasburger – Lehrbuch der Botanik. 35th ed. – Spektrum Akademischer Verlag, Heidelberg.

Skov, F. & J.-C. Svenning (2004): Potential impact of climatic change on the distribution of forest herbs in Europe. *Ecography* **27**: 366-380.

Svenning, J.-C. & F. Skov (2004): Limited filling of the potential range in European tree species. *Ecol. Letters* **7**: 565-573.

Sykes, M.T., Prentice, I.C. & W. Cramer (1996): A bioclimatic model for the potential distributions of north European tree species under present and future climates. *J. Biogeogr.* **23**: 203-233.

Thomas, C.D. & J.J. Lennon (1999): Birds extend their ranges northwards. *Nature* **399**: 213.

Thomas, C.D., Bodsworth, E.J., Wilson, R.J., Simmons, A.D., Davies, Z.G., Musche, M. & L. Conradt (2001): Ecological and evolutionary processes at expanding range margins. *Nature* **411**: 577-581.

Walter, H. & H. Straka (1970): Arealkunde – Floristisch-historische Geobotanik. 2nd ed. – Ulmer, Stuttgart.

Walther, G.-R. (2004): Plants in a warmer world. *Perspect. Plant Ecol. Evol. Syst.* **6**: 169-185.

Walther, G.-R., Burga, C.A. & P.J. Edwards (eds.) (2001): "Fingerprints" of Climate Change – Adapted behaviour and shifting species ranges. – Kluwer Academic / Plenum Publishers, New York.

Walther, G.-R., Post, E., Convey, P., Menzel, A., Parmesan, C., Beebee, T.J.C., Fromentin, J.-M., Hoegh-Guldberg, O. & F. Bairlein (2002): Ecological responses to recent climate change. *Nature* **416**: 389-395.

Woodward, F.I. (1987): Climate and plant distribution. – Cambridge University Press, Cambridge.

Woodward, F.I., Lomas, M.R. & C.K. Kelly (2004): Global climate and the distribution of plant biomes. *Phil. Trans. R. Soc. Lond. B* **359**: 1465-1476.

3 Distribution of evergreen broad-leaved woody species in Insubria in relation to bedrock and precipitation[2]

3.1 Abstract

An increasing number of evergreen broad-leaved species has been reported to naturalise in deciduous lowland forests near the lakes of southern Switzerland and northern Italy (Insubria), favoured by rising winter temperatures of the last few decades. Whereas existing studies on this issue concentrated on individual lakes or selected species only, the here presented survey covers the entire Insubrian region and focuses on differences in distribution of evergreen broad-leaved species and their local environmental constraints. Species' composition, cover values and degree of establishment of exotic and indigenous evergreen broad-leaved species were recorded on 22 study sites throughout the Insubrian region, from Lago d'Orta in the west to Lago di Garda in the east and analysed in relation to the regional precipitation gradient and differences in the geological bedrock. Whereas the general vegetation shift towards more evergreen broad-leaved species is consistent with a climate change explanation, distinct local differences in the distribution and composition of evergreen broad-leaved species within the area can be explained by the regional precipitation gradient and differences in the bedrock.

Keywords: Biogeography, climate change, exotic species, invasion, range margin, Ticino, vegetation shifts.

3.2 Introduction

The global distribution of evergreen broad-leaved vegetation is determined by climatic factors, with low winter temperatures and the length of the growing season as major factors limiting the range of these species towards the poles (e.g. BOX 1981, WALTER & BRECKLE 1999, WOODWARD et al. 2004). In Europe, repeated glaciations have diminished the diverse tertiary evergreen broad-leaved flora to a few species being able to recolonise central European forests in the course of the Holocene (SCHMID 1939, LANG 1994). Hence, there is now an impoverished pool of native species belonging to the hygrophil type of evergreen broad-leaved (= laurophyllous) vegetation in central Europe, compared to other regions with similar climatic conditions, i.e. in the

---

[2] Published in Botanica Helvetica.
Birkhäuser Verlag: www.springer.com/dal/home/birkhauser
Full reference: BERGER, S. & G.-R. WALTHER (2006): Distribution of evergreen broad-leaved woody species in Insubria in relation to bedrock and precipitation. *Bot. Helv.* **116**: 79-91.

transition zone between deciduous and evergreen broad-leaved forest in East Asia and North America (FUJIWARA & BOX 1994). However, in the past centuries, numerous exotic evergreen broad-leaved ornamentals were introduced from these areas and cultivated in gardens and parks in Europe wherever allowed by the local environmental conditions. These introduced species enriched the pool of warm-temperate/subtropical species for naturalisation in Europe considerably.

In recent years, an increasing number of introduced laurophyllous species has spread and naturalised in deciduous forests around the Insubrian lakes (GIANONI et al. 1988, KLÖTZLI et al. 1996, CARRARO et al. 1999, 2001, KLÖTZLI & WALTHER 1999, WALTHER 1999). Existing studies on this issue concentrated so far on individual lakes or selected species only, and, in some cases, only unpublished data or reports are available (LARCHER 1979, BRANDES 1989, BRULLO & GUARINO 1998, WALTHER et al. 2001, PROSSER 2002, WALTHER 2003, GUARINO & SGORBATI 2004, K. KÜTTEL unpubl. results, R. ZÄCH unpubl. results). The few existing studies on the flora and vegetation of the entire Insubrian region were published decades ago (SCHRÖTER 1936, SCHMID 1956, OBERDORFER 1964, PITSCHMANN et al. 1965), i.e. well before the period of pronounced expansion of evergreen broad-leaved species in the recent past (GIANONI et al. 1988, WALTHER 2001).

Previous studies have provided a better understanding of the preferences of these evergreen broad-leaved species in terms of their temperature requirements (SAKAI & LARCHER 1987, LARCHER 2000) and addressed the importance of rising winter temperatures in the last decades for their establishment (WALTHER 2000). However, knowledge of the species' specific preferences regarding other environmental factors is still lacking, but important to explain the spatial distribution of the species on a more local scale. In order to detect spatial patterns of the evergreen broad-leaved species' distribution, regarding the exotic laurophyllous species as well as other evergreen broad-leaved woody species, it is required to consider the Insubrian region as a whole, including its major environmental gradients. This study aims at compiling the knowledge from various local reports and complementing it with new field data. Forest stands rich in evergreen broad-leaved species were localised along the Insubrian lakes and sampled for their cover with evergreens. We compare the species composition and their degree of establishment in different study sites across the Insubrian region, in order to detect and explain spatial patterns in relation to a regional precipitation gradient and differences in the geological bedrock.

3.3 Study area

The study area is located around the larger Insubrian lakes from Lago d'Orta in the west to Lago di Garda in the east. It is covering the range from 8° 25' N to 10° 55' N and 45° 30' E to 46° 10' E, approximately (Fig. 3.1). For the purpose of this study only forest stands below 500 m a.s.l. were of interest, because the target species of this study are restricted to this altitudinal belt. Whereas siliceous substrates are dominant in the western part of the study area, the region of Lago di Garda and Lago d'Iseo is dominated by calcareous bedrock (Fig. 3.1 a).

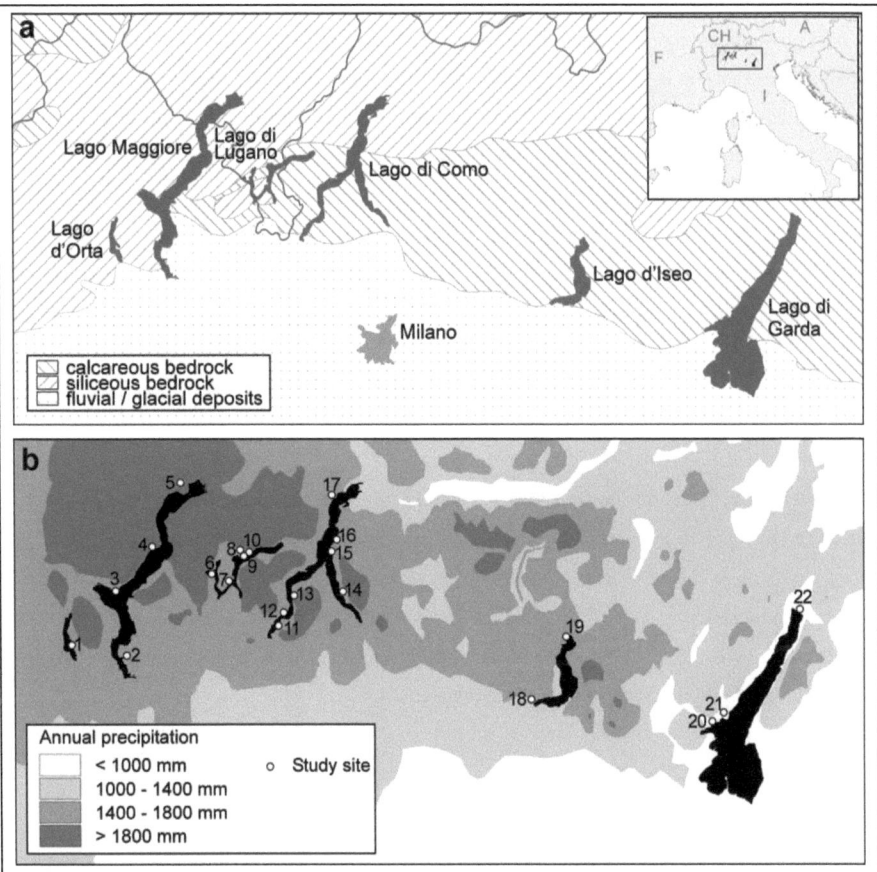

Fig. 3.1: Overview of the study area in the Insubrian lake region with information on a) bedrock, based on OBERDORFER (1964), modified, and b) the localities of the sample sites (for detailed information see Table 3.1) and annual precipitation (based on SCHWARB et al. 2000).

Insubria was early recognised as a distinct bioclimatic region, characterised by its mild climate (CHRIST 1879, SCHMID 1939, GIACOMINI & FENAROLI 1958). Within the region, a major climatic gradient from west to east is given by precipitation spanning a range from ca. 2000 mm per year in the region of Lago Maggiore in the west to less than 1000 mm per year at the southern end of Lago di Garda in the east (Fig. 3.1b). The effect of this gradient in the yearly amount of rainfall is further amplified by the seasonal pattern with summer rain in the west and 1-3 months of summer drought in the east (for details see REISIGL 1996).

The landscape at the southern foot of the Alps is characterised by the lakes and the adjacent steep slopes. The lowland forests near the lakes consist mainly of deciduous tree species (MAYER 1983). A predominant tree species in the study area is *Castanea sativa*, which was introduced by the Romans (CONEDERA et al. 2004). Among the most abundant native tree species are *Quercus* spp., *Tilia* spp., *Fraxinus* spp. and *Ostrya carpinifolia*. Only few indigenous evergreen broad-leaved species are able to grow up to the tree layer, such as *Quercus ilex, Ilex aquifolium, Laurus nobilis* (considered indigenous at least in the Italian part of the investigation area) as well as the liana *Hedera helix*. REISIGL (1996) distinguished two major types of lowland forest communities in western Insubria: Species-poor oak-birch forest on shallow soils, belonging to the alliance Quercion robori-petraeae. On mesic soils species-rich Insubrian deciduous hardwood forests are developed, belonging to the alliance Carpinion. Detailed floristic and phytosociological descriptions of the study area are provided by KNAPP (1953), OBERDORFER (1964), GIANONI et al. (1988), REISIGL (1996), BRULLO & GUARINO (1998), CARRARO et al. (1999) and WALTHER (2000).

In contrast to the deciduous forest vegetation, the ornamental garden and park vegetation around the lakes is rich in evergreen broad-leaved species originating from warmer temperate and subtropical areas, with for instance *Cinnamomum glanduliferum, Magnolia grandiflora, Camellia japonica, Rhododendron* spp., *Trachycarpus fortunei* and *Prunus laurocerasus* (e.g. SCHRÖTER 1936, SCHMID 1956). WALTHER (1999) gives a comprehensive list of the most abundant and often naturalising species in the Swiss part of the study area.

3.4 Methods

The probability to record neophytes with standardised monitoring protocols may be reduced due to the preference of these species to peculiar habitat types (e.g. proximity to anthropogenically influenced sites) (BÄUMLER et al. 2005). As most of the species investigated in this study are neophytes, we chose to focus on available adequate study sites, rather than using a regular grided sampling design. We based our selection of study sites on existing local surveys of forests rich in evergreen broad-leaved species in the region, as well as on interviews with local experts. Numerous recent vegetation records document laurophyllisation in various areas, such as near Lago Maggiore (CARRARO et al. 1999, WALTHER 2000, G. CARRARO unpubl. data), Lago di Lugano (K. KÜTTEL unpubl. data, R. ZÄCH unpubl. data) and in the area of Lago d'Iseo and Lago di Como (G. CARRARO unpubl. data). BRULLO & GUARINO (1998) provide an overview of the distribution of *Laurus nobilis*, *Ilex aquifolium* and *Quercus ilex* around Lago di Garda. Synanthropic species were also reported from Monte Brione (PROSSER 2002).

From these existing local surveys we selected the sites richest in evergreen broad-leaved species, which also represent the best developed stages of laurophyllisation in the region. The selection was based on the number of different evergreen broad-leaved species in the records and their cover values, where present. In addition, an overall survey of the area along the lakes was carried out in order to verify whether the potential sampling sites selected from previous work really represented the sites with the greatest share of evergreen broad-leaved species in the area. The survey was carried out in February/March 2005, as deciduous trees are still leafless in early spring and evergreen species thus easily recognised.

On each locality a sampling plot of the approximate size of 10 x 10 m was determined. The plot was locally positioned in a way to cover best all the occurring evergreen species in the particular site. Species composition and cover of evergreen broad-leaved species of each plot was recorded, using cover values of the Braun-Blanquet scale (BRAUN-BLANQUET 1964), separating the individuals in the herb layer (<0.5 m), in the shrub layer (>0.5 <7 m) and in the tree layer (>7 m). In total, we investigated 22 sample plots distributed over the study area.

In addition, the size of the tallest individuals of each species (up to a maximum of five plants, where present) at each locality was measured, or estimated if taller than 3 m. For the palm *Trachycarpus fortunei*, we measured the height of the woody stem, excluding the fronds. As a rule of thumb, *Trachycarpus fortunei* develops a woody stem after ca. 10 years under good growing conditions (cf. WALTHER 2003). As the petioles differ in length according to light availability, we regarded the height of the woody stem as a more robust measure for the growth of the individuals. Young specimens without woody stems were not included in the analysis of the size data of the tallest individuals. The size data of the most frequent species (*Laurus nobilis, Trachycarpus fortunei, Prunus laurocerasus* and *Cinnamomum glanduliferum*) was assigned to the respective precipitation class of their location (SCHWARB et al. 2000). The precipitation classes with less than 1400 mm/year were merged to one class. The average height and standard deviation of the sampled individuals of each species for each precipitation class was calculated separately for individuals growing on localities with calcareous bedrock and siliceous bedrock respectively.

Furthermore, the species' preferences, derived from the spatial pattern in their Insubrian distribution, were combined with and verified on the basis of ecological knowledge from the literature and the distribution of the species in their native habitat. This resulted in an ecogram for the seven most common evergreen broad-leaved species, which ranks their specific preferences regarding the gradient of precipitation and differences in the edaphic conditions in the Insubrian region.

The nomenclature of plant names follows GRIFFITHS (1994) and AESCHIMANN et. al. (2004).

3.5 Results

There are distinct differences in the composition, distribution and abundance of evergreen broad-leaved species between the study sites. Three biogeographic distribution patterns can be distinguished based on the data provided in Table 3.1: (i) Species occurring predominantly in the western part of the area (e.g. *Cinnamomum glanduliferum* and *Trachycarpus fortunei*), (ii) species prevailing in the eastern part of the area (e.g. *Quercus ilex*), and (iii) species consistent throughout the whole area, such as *Laurus nobilis*.

The distribution of the species within the study sites coincides with the regional precipitation gradient as well with differences in the geological conditions of the study area. *Cinnamomum glanduliferum* is restricted to the western part of the study area, where it was found exclusively on siliceous soils in areas with high precipitation (Table 3.1 and Fig. 3.2). The distribution of the palm *Trachycarpus fortunei* is also concentrated in the western part of the study area, in areas with high precipitation but without any particular edaphic preference (Table 3.1, cf. also Fig. 3.2). *Prunus laurocerasus* also shows high abundance in the sampling plots of the western part of the area, whereas there are only few scattered occurrences in the eastern part (Table 3.1).

In contrast, *Quercus ilex* was mainly found in the eastern part of the area, thus, in the sampling plots with lower precipitation and predominantly on calcareous soils. Also *Viburnum tinus* is restricted to the drier, calcareous sites (Table 3.1).

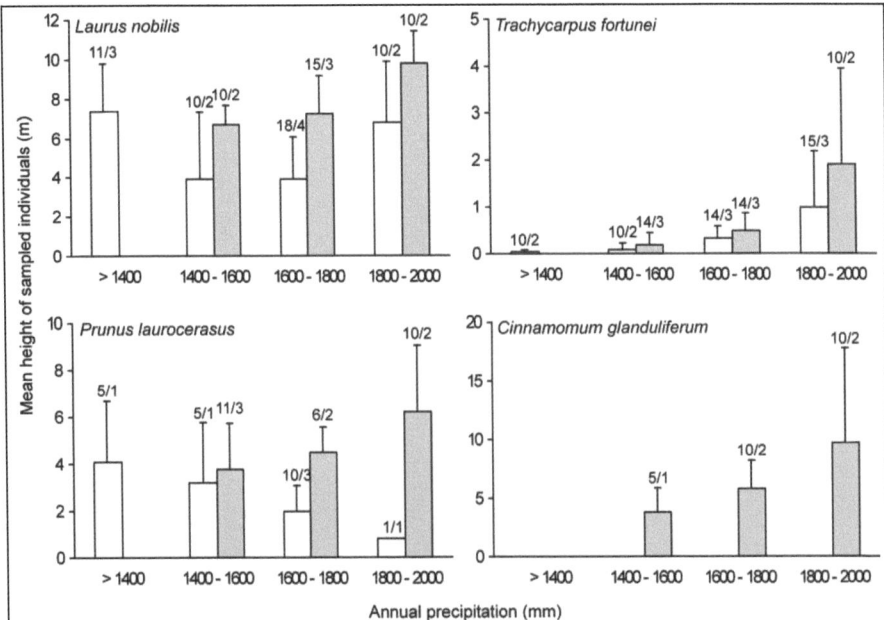

Fig. 3.2: The mean height and standard deviation of up to five sampled individuals per location of *Laurus nobilis*, *Trachycarpus fortunei*, *Prunus laurocerasus* and *Cinnamomum glanduliferum* in relation to the annual precipitation of their location according to SCHWARB et al. (2000). White bars show locations with calcareous bedrock, grey bars show locations with siliceous bedrock. Above each bar the number of sampled individuals/localities is indicated. The height of *Trachycarpus fortunei* is shown for the woody stem, excluding the fronds.

No clear preference for either end of the investigated gradient was found for *Ligustrum lucidum* and *Laurus nobilis*. The latter occurs consistently throughout the study area, with numerous specimens even present in the tree layer. Furthermore, the size of the tallest individuals of *Laurus nobilis* remains constant across all the investigated localities, independent of the varying precipitation regime (Fig. 3.2), contrary to the pattern shown by e.g. *Cinnamomum glanduliferum* (Fig. 3.2) and *Trachycarpus fortunei* (Fig. 3.2) with a clearly decreasing trend in size and frequency (Table 3.1), which is in parallel with decreasing precipitation values.

These peculiar spatial patterns of local species' distribution and analogies of regional differences of environmental conditions were synthesised in an ecogram showing the ecological preferences of the seven most common evergreen broad-leaved species regarding precipitation and bedrock in the investigated area (Fig. 3.3). The ecogram shows that *Cinnamomum glanduliferum* is restricted to the siliceous substrates whereas *Viburnum tinus* and *Quercus ilex* are occurring on calcareous bedrock. The other species occur on calcareous as well as on siliceous bedrock. Regarding the precipitation of their localities, the species are ranked from *Cinnamomum glanduliferum* and *Trachycarpus fortunei* on the upper end of the moisture gradient, followed by *Prunus laurocerasus* and *Ilex aquifolium* to *Laurus nobilis, Viburnum tinus* and *Quercus ilex* on the driest sites.

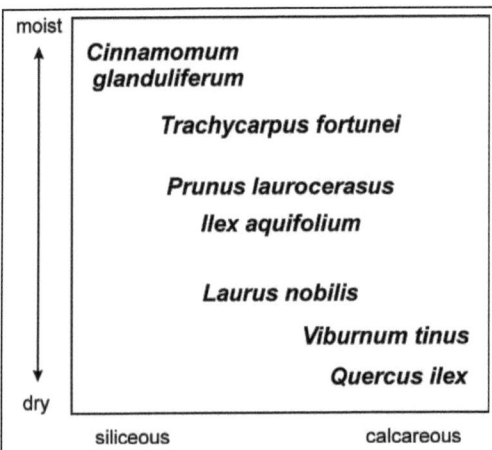

Fig. 3.3: Ecogram of the investigated species, ranking the species' specific preferences in the Insubrian area (cf. Table 3.1), regarding the bedrock (siliceous / indifferent / calcareous) and the precipitation gradient.

Table 3.1: Synoptic table of records with Braun-Blanquet cover values of evergreen broad-leaved species in the investigated plots. The records are sorted by soil type and decreasing precipitation. Prefixes indicates which layer the species refers to: H-: herb layer, S-: shrub layer and T-: tree layer.

Plot number	5	4	3	7	12	16	17	1	2	10	8	9	11	14	6	15	13	19	18	20	21	22
Location	Locarno-Solduno	Cannero	Verbania	Cannero di Fondo	Carate Urio	Bellano	Consiglio di Rumo	Orta	Angera	Gandria	Lugano	Castagnola	Blevio-Sopravilla	Mandrello del Lario	Monte Castellano	Varenna	Nesso	Lovere	Sarnico	Gardone-Rivera	Toscolano-Maderno	Monte Brione
Latitude (N)	46°10'22"	46°01'28"	45°55'24"	45°57'02"	45°52'24"	46°02'16"	46°06'33"	45°47'50"	45°47'01"	46°00'29"	46°00'59"	46°00'14"	45°50'07"	45°55'41"	45°57'40"	46°00'38"	45°54'48"	45°49'17"	45°40'30"	45°37'19"	45°39'57"	45°52'46"
Longitude (E)	08°46'56"	08°40'54"	08°33'42"	08°56'29"	09°06'59"	09°18'06"	09°17'14"	08°24'41"	08°34'08"	09°00'19"	08°58'28"	08°59'28"	09°06'08"	09°19'25"	08°52'53"	09°17'15"	09°09'33"	10°04'26"	09°57'07"	10°33'24"	10°37'27"	10°51'49"
Lake	Maggiore	Maggiore	Maggiore	Lugano	Como	Como	Como	Orta	Maggiore	Lugano	Lugano	Lugano	Como	Como	Lugano	Como	Como	Iseo	Iseo	Garda	Garda	Garda
Date	02.03.05	02.03.05	02.03.05	06.03.05	01.03.05	01.03.05	01.03.05	08.03.05	08.03.05	28.02.05	28.02.05	09.03.05	05.03.05	10.03.05	09.03.05	10.03.05	05.03.05	03.03.05	03.03.05	07.03.05	07.03.05	07.03.05
Altitude (m a.s.l.)	240	300	200	300	250	210	250	330	250	340	350	370	420	300	280	220	350	250	260	75	500	80
Aspect	S	SSO	E	SSE	SE	WNW	SE	ENE	E	SSE	NW	SW	NW	ESE	SE	SSW	WNW	-	S	S	SW	WNW
Inclination (%)	30	70	5	30	80	20	50	35	35	40	60	30	-	30	10	30	30	-	30	50	25	25
Bedrock (sil-siliceous, ca-calcareous)	si	si	si	si	si	si	si	si	-	ca	ca	ca	ca	ca	ca	ca	ca	ca	ca	ca	ca	ca
Annual precipitation (mm)	1800-2000	1800-2000	1600-1800	1600-1800	1600-1800	1400-1600	1400-1600	1400-1600	1400-1600	1800-2000	1800-2000	1800-2000	1600-1800	1600-1800	1600-1800	1600-1800	1400-1600	1400-1600	1300-1400	1000-1100	900-1000	800-900
T-Cinnamomum glandulifferum	+																					
H-Cinnamomum glandulifferum	+		+																			
S-Cinnamomum glandulifferum	1	+		+		+																
S-Aucuba japonica						r																
T-Ligustrum lucidum			+																			
T-Prunus laurocerasus	2	+		+	1			r														
S-Prunus laurocerasus	2	2	2	+		2		1		r												
H-Prunus laurocerasus	1	1	+			1		r														
S-Trachycarpus fortunei	3		+	+	1		+	+	1		3		1	+	r		1					+
H-Trachycarpus fortunei	1	+	+	+			+	+	2	1	1	+		+	r		+	r	r	2		1
S-Laurus nobilis	1	2	2	+	1	1	1		2	+	1	+	2	+	+	1	1	r	r	1		2
H-Laurus nobilis	+	1	1	1		1	1		1	+	1	+	1	+	+	+	2	r		+		1
T-Laurus nobilis	1				2		2			2			1	+	+	1	1		r	1		
S-Elaeagnus pungens	1							1						+		+	r		+	+	+	1
S-Ilex aquifolium	+		+					+	r					+	r		r				1	
H-Ilex aquifolium	+		+					+	r					+			+					
T-Ilex aquifolium								r														
S-Ligustrum lucidum		+			+									+	+			1	1	+		1
H-Ligustrum lucidum														+	+			1	+			1
S-Viburnum tinus											+				+	1			1	+		1
H-Viburnum tinus												+			+	+			+	+		+
S-Quercus ilex															+			1	1			
T-Quercus ilex															+				+		+	+
H-Quercus ilex															+		r		+		+	+
H-Pittosporum tobira																+	r			+		+
S-Pittosporum tobira																+						+

Other evergreen broadleaved species (plot number: layer-species cover value): 3: S-*Mahonia* cf. *bealei* +, 6: H-*Elaeagnus pungens* r, 9: S-*Pyracantha coccinea* +, 22: S-*Eriobotrya japonica* +.

31

3.6 Discussion

So far, the investigation of the evergreen broad-leaved vegetation has mainly focused on the effects of rising winter temperature of the recent past. It was shown that important temperature thresholds for the establishment and growth of the species representing this vegetation type were exceeded only a few decades ago, due to recent climate change (CARRARO *et al.* 1999, WALTHER 2002a, WALTHER *et al.* 2002).

The minor differences in temperature within the region do not explain the observed differences within the set of target species, because they allow the survival of all the evergreen broad-leaved species on favourable sites on southern slopes close to every Insubrian lake. Furthermore, especially the more frequently escaped species (cf. Table 3.1) are so widely cultivated in the entire area that the pattern of their subspontaneous population is not biased substantially by the availability of seed sources in gardens and parks. Hence, other environmental variables are needed to explain the obvious preference of some of the evergreen broad-leaved species for peculiar sites within the investigated area. Distinct regional environmental gradients from west to east are given by the amount and regime of yearly precipitation, and by differences in the geological substrate, with predominantly siliceous soils in the west and calcareous soils in the east. These parameters are known to be important ecological factors limiting the distribution of individual species on regional to local scales (e.g. SITTE *et al.* 2002, POTT 2005), and are hereafter outlined with regard to the origin, ecology and physiology of the most frequently occurring evergreen broad-leaved species.

Cinnamomum glanduliferum was exclusively found on siliceous soils in the western area with high precipitation. The native range of the genus as well as morphological features of this particular tree species (e.g. the acuminate leaf apex) indicate its affinity to moist subtropical conditions (cf. RAVINDRAN *et al.* 2004). The latter allow fast growth of the young plants when water supply and nutrients are not limited. In the area of the southern foot of the Alps, this is only the case in the parts with high precipitation and siliceous bedrock, which allow the accumulation of humus in the top soil layer.

Trachycarpus fortunei occurs regularly and with high cover values on sites in the western part of the area, whereas the occurrence in the eastern part is restricted to very few localities with low densities (Table 3.1, cf. also Fig. 3.2). This pronounced decline in frequency and the decreasing size of the *Trachycarpus fortunei* individuals are in line with the decrease in precipitation towards the east. This indicates that the species is struggling to establish when approaching its precipitation limit. The easternmost study sites are characterised by more pronounced summer droughts which may limit the growth of this subtropical species, also supported by the fact that *Trachycarpus fortunei* is restricted to areas with a humid climate in its native range (WALTHER 2002b). Thus, in the eastern part of the investigation area *Trachycarpus fortunei* occurs in very few places, and, in contrast to the established subspontaneous populations in the western part, there is still doubt about the persistence of the individuals found in the eastern part (see PROSSER 2002).

Prunus laurocerasus also shows high abundance in the sampling plots in the western part of the area, whereas the occurrences are only scattered in the eastern part. This underlines the affiliation of the species to the western vegetation type, also referred to as *Laurocerasus*-belt (see SCHMID 1956). In the native range, *Prunus laurocerasus* is known to grow on both calcareous and siliceous soils in areas with annual precipitation above 1200 mm (ZAZANASHVILI 1999). In the Turkish part of the Black Sea region, *Prunus laurocerasus* occurs in areas with an annual mean temperature of 14.5 °C where the annual rainfall, distributed throughout the year, exceeds 830 mm (ISLAM 2002). The distribution pattern in Insubria and the information from the native range (cf. also Fageta laurocerasosa in NAKHUTSRISHVILI 1999) suggest that climatic parameters in the study area, especially the increasing risk of drought periods towards the east, have a limiting effect on the distribution of *Prunus laurocerasus*.

In contrast, *Laurus nobilis* is present in almost all the sampled plots, independent of the geological substrate. This is in agreement with information from horticultural literature (KRÜSSMANN 1960) and natural stands of *Laurus nobilis* throughout the Mediterranean region (cf. BRULLO et al. 2001). Despite its broader amplitude in terms of water supply, *Laurus nobilis* was ranked on the drier end of the gradient in the ecogram (Fig 3.3) to underline its relative drought tolerance compared to the other aforementioned evergreen broad-leaved species. Consequently, except from temperature, *Laurus nobilis* meets no major limitations within the investigated area, neither regarding the precipitation regime nor in terms of the geological conditions.

A species mainly found in the eastern part of the area is the Mediterranean *Quercus ilex*. The only exclave in the western part of the study area, where *Quercus ilex* regenerates subspontaneously, is on the south-facing slope of Monte Caslano. This site is clearly distinguished from the surrounding area by its calcareous, shallow soils with reduced water storage capacity and consequent temporary drought stress, despite relatively high annual precipitation (cf. COTTI et al. 1990, ZÄCH unpubl.). In the entire region, this relatively slow growing sclerophyllous species profit on shallow calcareous soils from reduced competition from either deciduous species (LARCHER 1979, BARBERO et al. 1992, BRULLO & GUARINO 1998, cf. also MITRAKOS 1980, KOLLMANN & PFLUGSHAUPT 2005), or other, faster growing evergreen broad-leaved (laurophyllous) species. This explains the restriction of this species at its northern fringe of distribution to drier sites with calcareous soils, whereas in the core area of distribution no clear soil preference is expressed by *Quercus ilex* (S. SABATE, pers com.). The same pattern is shown for *Viburnum tinus* which is often associated with *Quercus ilex* (see also BARBERO *et al.* 1992). It shares the same ecological preferences for dry, calcareous sites in the here investigated area, in spite of its edaphic indifference in eumediterranean areas of its distribution (see TORRES *et al.* 2002).

Whereas the aforementioned species occur sufficiently frequently in the area to derive information on their ecological preferences, others (cf. Table 3.1) are still restricted by the varying availability of seed sources in gardens and parks. Hence, the restriction of *Aucuba japonica* to the western part, and the opposite for *Pittosporum tobira* seems rather to be an artefact of sample size, and is lacking confirmation from the literature. For instance, *Pittosporum tobira* has also been reported to grow on siliceous soils in the western part of the area (WALTHER 1999). In its native habitat, *Pittosporum tobira* grows on dry sites with shallow soils, such as mountain ridges, indicating some extent of drought resistance (NAKAMURA et al. 2000).

In general, this study shows that the effect of the precipitation gradient in the study area is modified by the geological bedrock with clear effects on the distribution of sensitive species. In the western part, precipitation is high, and there is mainly siliceous bedrock. The decay of organic matter is relatively slow and allows the accumulation of a relatively thick humus layer (BLASER 1973). In contrast, the calcareous underground in the eastern part of the area provides soils with a lower water storage capacity, increasing the risk of drought stress. With increasing risk of drought stress the laurophyllous species become less abundant and they are increasingly replaced by sclerophyllous, (sub-)Mediterranean species. Hence, not only the native vegetation, but also the distribution and composition of introduced evergreen broad-leaved species reflect the two bioclimatic and floristic Insubrian subtypes distinguished in the literature (e.g. OBERDORFER 1964, REISIGL 1996).

THOMAS *et al.* (2001) emphasised that habitat requirements of species become more specific towards range boundaries. Hence, the peripheral areas of species' distribution may provide best estimates for a detailed understanding of species' specific ecological preferences, which is of current interest, as species at their northern boundaries have recently expanded their habitat ranges as conditions have become less marginal, due to climate change (THOMAS *et al.* 2001, WALTHER *et al.* 2005, see also WALTHER 2004).

3.7 Conclusions

Whereas the general vegetation shift towards more evergreen broad-leaved species is consistent with a climate change explanation, distinct differences on the species level suggest further environmental constraints resulting in the varying composition, abundance and frequency of evergreen broad-leaved species within the area. Evergreen broad-leaved species were detected in lowland deciduous forests near all the lakes, although the forested area and thus the availability of suitable investigation sites decreases towards the east. The distinct distribution patterns of the different species and their varying ability to establish in the studied sites is related to the regional precipitation gradient and differences in the bedrock.

3.8 Acknowledgements

The authors thank G. Carraro (DIONEA SA., Locarno) and R. Guarino (Cagliari) for their contribution of local data and expertise and assistance during the field work. The study was funded by the EC within the FP6 Integrated Project 'ALARM'; GOCE-CT-2003-506675.

3.9 Zusammenfassung

Berger. S. und Walther G.-R. Verbreitung von immergrünen Laubholzarten in Insubrien in Bezug zum geologischen Untergrund und Niederschlag.

Zahlreiche immergrüne Laubholzarten haben sich, begünstigt durch die angestiegenen Wintertemperaturen der letzten Jahrzehnte, in den sommergrünen Tieflagenwäldern um die großen Seen der Südschweiz und Norditaliens (Insubrien) etablieren können. Während bisherige Untersuchungen sich auf lokale Gebiete oder bestimmte Arten beschränkten, umfasst diese Studie die gesamte Insubrische Region und analysiert die Unterschiede in der Verbreitung der einzelnen immergrünen Arten und im Artenspektrum der untersuchten Standorte, sowie deren Zusammenhang mit auf lokaler Ebene maßgeblichen Umweltfaktoren. Artenspektrum, Deckungsgrade und der Etablierungsgrad der immergrünen Laubholzarten wurde an 22 Untersuchungspunkten zwischen Orta- und Gardasee erfasst und in Bezug zum regionalen Niederschlagsgradienten und zu geologischen Unterschieden gestellt. Während die allgemeine Zunahme an immergrünen Arten mit dem Klimawandel im Zusammenhang steht, können lokale Unterschiede in der Verbreitung der einzelnen Arten innerhalb der Region und im Artenspektrum der untersuchten Standorte anhand des regionalen Niederschlagsgradienten und Unterschieden im geologischen Untergrund erklärt werden.

3.10 References

AESCHIMANN, D., LAUBER, K., MOSER, D.M. & J.P. THEURILLAT (2004): Flora alpina. – Haupt, Bern.

BARBERO, M., LOISEL, R. & P. QUEZEL (1992): Biogeography, ecology and history of Mediterranean *Quercus ilex*-ecosystems. *Vegetatio* **100**: 19-34.

BÄUMLER, B., MOSER, D.M., GYGAX, A., LATOUR, C. & N.WYLER (2005): Fortschritte in der Floristik der Schweizer Flora (Gefäßpflanzen). *Bot. Helv.* **115**: 83-93.

BLASER, P. (1973): Die Bodenbildung auf Silikatgestein im südlichen Tessin. *Mitt. Schweiz. Anst. forstl. Vers.wes.* **49**: 254-340.

BOX, E.O. (1981): Macroclimate and plant forms: An introduction to predictive modelling in phytogeography. Tasks for vegetation science 1. – Junk Publishers, The Hague.

BRANDES, D. (1989): Zur Soziologie einiger Neophyten des insubrischen Gebietes. *Tuexenia* **9**: 267-274.

BRAUN-BLANQUET, J. (1964): Pflanzensoziologie. 3. ed. – Springer, Wien.

BRULLO, S. & R. GUARINO (1998): The forest vegetation from the Garda lake (N Italy). *Phytocoenologia* **28**: 319-355.

BRULLO, S., COSTANZO, E. & V. TOMASELLI (2001): Phytosociological study on the *Laurus nobilis* communities in the Hyblaean Mountains (SE Sicily). *Phytocoenologia* **31**: 249-270.

CARRARO, G., KLÖTZLI, F., WALTHER, G.-R., GIANONI, P. & R. MOSSI (1999): Observed changes in vegetation in relation to climate warming. Final Report NRP 31. – vdf Hochschulverlag, Zürich.

CARRARO, G., GIANONI, G., MOSSI, R., KLÖTZLI, F. & G.-R. WALTHER (2001): Observed changes in vegetation in relation to climate warming. In: BURGA, C.A. & KRATOCHWIL, A. (eds.): Biomonitoring: General and applied aspects on regional and global scales. Tasks for vegetation science 35. pp. 195-205. – Kluwer academic publishers, Dordrecht.

CHRIST, H. (1879): Das Pflanzenleben der Schweiz. – Bern, Zürich.

CONEDERA, M., KREBS, P., TINNER, W., PRADELLA, M. & D. TORRIANI (2004): The cultivation of *Castanea sativa* (Mill.) in Europe, from its origin to its diffusion on a continental scale. *Veg. Hist. Archeobot.* **13**: 161-179.

COTTI, G., FELBER, M., FOSSATI, A., LUCCHINI, G., STEIGER, E. & P.L. ZANON (1990): Introduzione al paesaggio naturale del Cantone Ticino. 1. Le componenti naturali. – Armando Dado editore, Locarno.

FUJIWARA, K. & E.O. BOX (1994): Evergreen broad-leaved forests of the southeastern United States. In: MIYAWAKI, A., IWATSUKI, K. & M.M. GRANDTNER (eds.): Vegetation in Eastern North America. pp. 273-312. – University of Tokyo Press, Tokyo.

GIACOMINI, V. & L. FENAROLI (1958): La Flora. Conosci l'Italia 2. – Milano.

GIANONI, G., CARRARO, G. & F. KLÖTZLI (1988): Thermophile, an laurophyllen Pflanzenarten reiche Waldgesellschaften im hyperinsubrischen Seenbereich des Tessins. *Ber. Geobot. Inst. ETH, Stiftung Rübel, Zürich* **54**: 164-180.

GRIFFITHS, M. (1994): Index of garden plants. The New Royal Horticultural Society Dictionary. – Macmillan, London.

GUARINO, R. & S. SGORBATI (2004): Guida Botanica al parco Alto Garda Bresciana. –Tipolitografia Bongi di Altini Paolo & Luigi snc, San Miniato.

ISLAM, A. (2002): 'Kiraz' cherry laurel (*Prunus laurocerasus*). *N. Z. J. Crop Hortic. Sci.* **30**: 301-302.

KLÖTZLI, F., WALTHER, G.-R., CARRARO, G. & A. GRUNDMANN (1996): Anlaufender Biomwandel in Insubrien. *Verh. Ges. Ökol.* **26**: 537-550.

KLÖTZLI, F. & G.-R. WALTHER (1999): Recent vegetation shifts in Switzerland. In: KLÖTZLI, F. & G.-R WALTHER (eds.): Conference on recent shifts in vegetation boundaries of deciduous forests, especially due to general global warming. pp. 273-300. – Birkhäuser, Basel.

KNAPP, R. (1953): Studien zur Vegetation und pflanzengeographischen Gliederung Nordwest-Italiens und der Süd-Schweiz. *Kölner geogr. Arb.* **4**: 5-59.

KOLLMANN, J. & K. PFLUGSHAUPT. (2005): Population structure of a fleshy-fruited species at its range edge – the case of *Prunus mahaleb* L. in northern Switzerland. *Bot. Helv.* **115**: 49-61.

KRÜSSMANN, G. (1960): Handbuch der Laubgehölze. – Paul Parey, Berlin.

LANG, G. (1994): Quartäre Vegetationsgeschichte Europas. – Gustav Fischer, Jena.

LARCHER, W. (1979): Climate and plant life of Arco. – Azienda autonoma di cura e soggiorno, Arco.

LARCHER, W. (2000): Temperature stress and survival ability of Mediterranean sclerophyllous plants. *Plant Biosystems* **134**: 279-295.

MAYER, H. (1983): Waldgebiete der Alpen. *Tüxenia* **3**: 307-318.

MITRAKOS, K. (1980): A theory for Mediterranean plant life. *Acta Oecologica-Oecologica Plantarum* **1**: 245-252.

NAKAMURA, Y., DE LA TORRE, W.W., DEL-ARCO AGUILAR, M.J. & J.A. REYES-BETANCORT (2000): A phytosociological study on Mediterranean laurel forest area of Tenerife, Canary Islands - in comparison with Japanese laurel forest landscape area of Izu, Central Japan. *Phytocoenologia* **30**: 613-632.

NAKHUTSRISHVILI, G. (1999): The vegetation of Georgia (Caucasus). *Braun-Blanquetia* **15**: 1-74.

OBERDORFER, E. (1964): Der insubrische Vegetationskomplex, seine Struktur und Abgrenzung gegen die submediterrane Vegetation in Oberitalien und in der Südschweiz. *Beitr. naturk. Forsch. SW-Deutschl.* **23**: 141-187.

PITSCHMANN, H., REISIGL, H. & H. SCHIECHTL (1965): Flora der Südalpen. – Gustav Fischer, Stuttgart.

POTT, R. (2005): Allgemeine Geobotanik. – Springer, Berlin.

PROSSER, F. (2002): Flora of Monte Brione by Riva del Garda (district of Trento). *Atti Acc. Rov. Agiati* **252** (ser VIII, vol. II B): 211-312.

RAVINDRAN, P.N., NIRMAL BABU, K. & M. SHYLAJA (eds.) (2004): Cinnamon and Cassia -The genus *Cinnamomum*. – CRC Press, Boca Raton.

REISIGL, H. (1996): Insubrien und das Gardaseegebiet - Vegetation, Florengeschichte, Endemismus. *Ann. Mus. civ. Rovereto* **11** (Suppl. II): 9-25.

SAKAI, A. & W. LARCHER (1987): Frost survival of plants. Ecological studies 62. – Springer, Berlin.

SCHMID, E. (1939): Die Stellung Insubriens im Alpenbereich. *Verh. Schweiz. Nat.forsch. Ges.*: 64-65.

SCHMID, E. (1956): Flora des Südens. – Rascher, Zürich.

SCHRÖTER, C. (1936): Flora des Südens. – Rascher, Zürich.

SCHWARB, M., DALY, C., FREI, C. & C. SCHÄR (2000): Mittlere jährliche Niederschlagshöhen im europäischen Alpenraum 1971-1990. – Hydrologischer Atlas der Schweiz, Bern.

SITTE, P., WEILER, E.W., KADEREIT, J.W., BRESINSKY, A. & C. KÖRNER (2002): Strasburger Lehrbuch der Botanik. – Spektrum, Heidelberg.

THOMAS, C.D., BODSWORTH, E.J., WILSON, R.J., SIMMONS, A.D., DAVIES, Z.G., MUSCHE, M. &, L. CONRADT (2001): Ecological and evolutionary processes at expanding range margins. *Nature* **411**: 577-581.

TORRES, J.A., VALLE, F., PINTO, C., GARCIA-FUENTES, A., SALAZAR, C. & E. CANO (2002): *Arbutus unedo* L. communities in southern Iberian Peninsula mountains. *Plant Ecol.* **160**: 207-223.

WALTER, H. & S.-W. BRECKLE (1999). Vegetation und Klimazonen. 7. edn. – Ulmer, Stuttgart.

WALTHER, G.-R. (1999): Distribution and limits of evergreen broad-leaved (laurophyllous) species in Switzerland. *Bot. Helv.* **109**: 153-167.

WALTHER, G.-R. (2000): Climatic forcing on the dispersal of exotic species. *Phytocoenologia* **30**: 409-430.

WALTHER, G.-R. (2001): Laurophyllisation - a sign of a changing climate? In: BURGA, C.A. & A. KRATOCHWIL (eds.): Biomonitoring: General and applied aspects on regional and global scales. Tasks for vegetation science 35. pp. 207-223. – Kluwer Academic Publishers, Dordrecht.

WALTHER, G.-R., CARRARO, G. & F. KLÖTZLI (2001): Evergreen broad-leaved species as indicators for climate change. In: WALTHER, G.-R., BURGA, C.A. & P.J. EDWARDS, (eds.): "Fingerprints" of Climate Change – Adapted behaviour and shifting species ranges. pp. 151-162. – Kluwer Academic/Plenum Publishers, New York.

WALTHER, G.-R. (2002a): Weakening of climatic constraints with global warming and its consequences for evergreen broad-leaved species. *Folia Geobot.* **37**: 129-139.

WALTHER G.-R. (2002b): Die Verbreitung der Hanfpalme *Trachycarpus fortunei* im Tessin – 50 Jahre nach der Erstaufnahme. *Schweiz. Beitr. Dendrol.* **47**: 29-41.

WALTHER, G.-R., POST, E., CONVEY, P., MENZEL, A., PARMESAN, C., BEEBEE, T.J.C., FROMENTIN, J.M., HOEGH-GULDBERG, O. & F. BAIRLEIN (2002): Ecological responses to recent climate change. *Nature* **416**: 389-395.

WALTHER, G.-R. (2003): Are there indigenous palms in Switzerland? *Bot. Helv.* **113**: 159-180.

WALTHER, G.-R. (2004): Plants in a warmer world. *Perspect. Plant Ecol. Evol. Syst.* **6**: 169-185.

WALTHER, G.-R., BERGER, S. & M.T. SYKES (2005): An ecological 'footprint' of climate change. *Proc. R. Soc. Lond.* B **272**: 1427-1432.

WOODWARD, F.I., LOMAS, M.R. & C.K. KELLY (2004): Global climate and the distribution of plant biomes. *Phil. Trans. R. Soc. Lond. B* **359**: 1465-1476.

ZAZANASHVILI, N. (1999): On the Colkhic vegetation. In: KLÖTZLI, F. & G.-R. WALTHER (eds.). Conference on recent shifts in vegetation boundaries of deciduous forests, especially due to general global warming. pp. 181-197. – Birkhäuser, Basel.

4 Palms tracking climate change[3]

4.1 Abstract

Aim: Many species are currently expanding their ranges in response to climate change but the mechanisms underlying these range expansion are in many cases poorly understood. In this paper we explore potential climatic factors governing the recent establishment of new palm populations far north of any other viable palm population in the world.

Location: Southern Switzerland, Europe, Asia, and the world.

Methods: We identified ecological threshold values for the target species based on gridded climate data, altitude and distributional records from the native range and applied them to the introduced range using local field monitoring and measured meteorological data as well as a bioclimatic model.

Results: We identify a strong relationship between minimum winter temperatures, influenced by growing season length and the distribution of the palm in its native range. Recent climate change strongly coincides with the palms' recent spread into southern Switzerland, which is in concert with the expansion of the global range of palms across various continents.

Main conclusions: Our results strongly suggest that the expansion of palms into (semi-) natural forests is driven by changes in winter temperature and growing season length and not by delayed population expansion. This implies that this rapid expansion is likely to continue in the future under a warming climate. Palms in general, and *Trachycarpus fortunei* in particular, are significant bioindicators across continents for contemporary climate change and reflect a global signal towards warmer conditions.

Key words: *Trachycarpus fortunei*; Arecaceae; global warming; invasion; vegetation shift; exotic species; bioindicator; biogeography, northernmost palm population, Ticino.

---

[3] Published in Global Ecology and Biogeography.
Blackwell publishing: http://www.blackwellpublishing.com/
Full reference: WALTHER, G.-R., GRITTI, E.S., BERGER, S., HICKLER, T., TANG, Z. & M.T. SYKES (2007): Palms tracking climate change. *Global. Ecol. Biogeogr.* **16**: 801-809.

4.2 Introduction

In recent years and as a result of recent climate warming, changes in the behaviour and distribution of species, the composition of and interactions within communities, and the structure and dynamics of ecosystems have been observed in an array of habitats from the poles to equatorial regions (e.g. HUGHES 2000, WALTHER *et al.* 2002, PARMESAN & YOHE 2003, ROOT *et al.* 2003). Plants for example, are responding to the enhanced warming of recent decades by changing their phenological patterns and/or shifting their ranges to higher latitudes or altitudes (WALTHER 2004). Furthermore, evidence is arising that not only indigenous species are responding to but also introduced species may profit from changing environmental conditions (WALTHER 2000, SOBRINO *et al.* 2001, see also DUKES & MOONEY 1999, SIMBERLOFF 2000). At the southern foot of the Alps, among the assemblage of invading evergreen broad-leaved plants, an introduced palm species (*Trachycarpus fortunei* (Hook.) Wendl.) has successfully colonised deciduous forests and established a vigorous population that already has fertile individuals (GIANONI *et al.* 1988, CARRARO *et al.* 1999, WALTHER 2003).

In the palaeobotanical literature, palms in general are recognised as effective bioindicators of warm climates. The presence of fossil palm remains in the geological record is invariably interpreted as indicative of warm and humid climatic conditions during the formation of the particular geological stratum (e.g. MAI 1995).

Under current global climates, palms reach their greatest proliferation in the tropics and are much less prominent and diverse in temperate regions (GOOD 1953, JONES 1995, GIBBONS 2003, LÖTSCHERT 2006). However, in recent years, evidence is increasing that the most cold-hardy palm species are occurring beyond the usual latitudinal range limit of palms (STÄHLER 2000, WALTHER 2002a, FRANCKO 2003). We here compile and synthesise the various recent records reporting new occurrences of palms outside the known range of global palm distribution, focusing on the hemp palm, *Trachycarpus fortunei*, the most widely cultivated species at the latitudinal palm range margin. We explore potential climatic factors governing the recent establishment of a new subspontaneous palm population in southern Switzerland, far north of any other viable palm population in the world. Based on gridded climate data, altitude and distributional records from China, i.e. the native habitat of *Trachycarpus fortunei*, we aim at identifying the limiting climatic parameters in the native range and verifying whether a shift in climate might explain the palms' recent spread in the introduced range south of the European Alps using a bioclimatic model. These findings are then applied to the global scale in order to assess whether there is an observed coherent range expansion of this species on various continents, which can then be interpreted as a global signal of the shifting in climate towards warmer conditions.

4.3 Material and Methods

The native habitat of *Trachycarpus fortunei* is located in south-eastern Asia (DELECTIS FLORAE REIPUBLICAE POPULARIS SINICAE 1991, WU & DING 1999, GIBBONS 2003) as shown in Fig. 4.1.

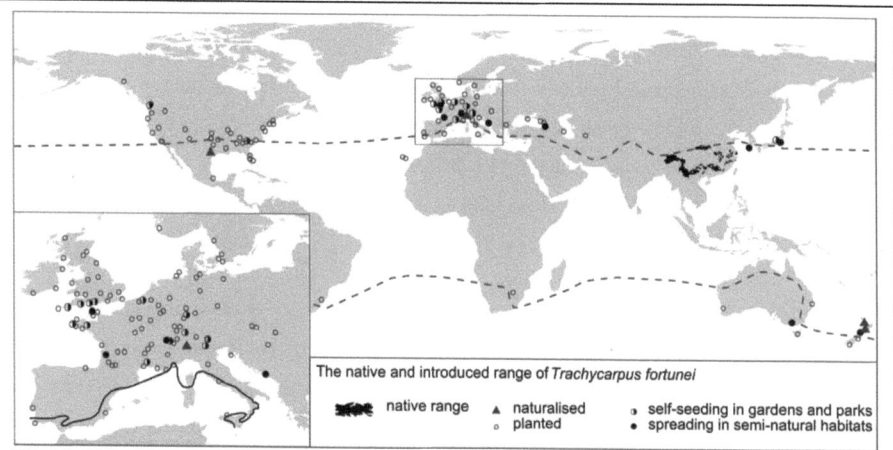

Fig. 4.1: The native range of *Trachycarpus fortunei* in China and a compilation of sites where *Trachycarpus fortunei* occurs outside its native range. The broken lines indicate the poleward range margins of global natural palm distribution. In the inset, the solid line shows the northern limit of the former northernmost palm population (*Chamaerops humilis*). The symbols represent a (non-exhaustive) compilation of *Trachycarpus*-sites based upon literature and internet search as well as personal observations and contacts; for details see text, acknowledgements and appendix 3.

Based on the distribution of this particular palm species in its native range (DELECTIS FLORAE REIPUBLICAE POPULARIS SINICAE 1991, WU & DING 1999) we derived the species' specific climatic requirements with regard to temperature and water availability. Whereas the absolute value of low temperatures limiting the species' survival have intensively been studied (WINTER 1976, LARCHER & WINTER 1981, SAKAI & LARCHER 1987, cf. also WALTHER 2002a), the experience from field experiments suggests that at least two climatic factors, i.e. low temperatures in winter and length of the growing season, are involved in limiting this species' northern/upper distribution when exposed to natural conditions in the field (WALTHER 2003, see also FITZROYA 2004).

In this study, we applied two approaches for deriving climatic constraints to the species' distribution in its native range. Both approaches used monthly climate data from the CRU dataset (0.5°; NEW et al. 2000) averaged over the period 1961-1990.

Firstly, we superimposed gridded data of minimum monthly temperatures and growing season length (growing degree days above 5°C (GDD_5) per year) on the distribution map of *Trachycarpus fortunei* in its native range in China in order to define the species' limits with regard to these two bioclimatic variables. Information on local altitudinal limits (DELECTIS FLORAE REIPUBLICAE POPULARIS SINICAE 1991, WU & DING 1999) was used to exclude those grid cells within the range of the species but where the reference altitude was higher than the known upper elevational limit of locally occurring *Trachycarpus*. The resulting two subsets of grid cells separate grid cells whose geographical position (latitude/longitude) or reference altitude in the gridded climate data set is outside the range of *Trachycarpus fortunei* occurrences from those which overlap with the range of *Trachycarpus fortunei*. The difference in climate between these two subsets was used to define the potential range especially towards colder areas (cf. BEERLING et al. 1995, EDWARDS et al. 1998, GUISAN & THUILLER 2005).

Secondly, although we here focus on the northern/upper limit of palm ranges, an estimate of the complete bioclimatic envelope of this species was derived by comparing the realised and the modelled distribution in the native range, using the bioclimatic model STASH (cf. SYKES et al. 1996), based on gridded climate data derived from the CRU dataset for the period 1961-1990. This temporal envelope compares well to the available plant distribution data on its native habitat (DELECTIS FLORAE REIPUBLICAE POPULARIS SINICAE 1991). The bioclimatic model (STAtic SHell) is a simple model which uses a minimum set of bioclimatic parameters (mean temperature of the coldest month, growing season warmth, a drought index and for some species a requirement for chilling before budburst in the spring) to describe a species range. These parameters are assumed to present responses to important physiologically mechanisms within a plant, for example growing season length (growing degree days) is an index of the presence of energy suitable for the completion of a plant's life cycle. Some of these parameters act as on-off switches, if for example, the minimum mean coldest month temperature (which is a surrogate for the absolute minimum (PRENTICE et al. 1992)) in a grid cell falls below the species' minimum limit, that species is excluded from that cell. Other parameters act directly upon net assimilation and respiration and thus on growth rate which is reflected as the degree of establishment success in a grid cell (SYKES et al. 1996 for full details).

For the introduced range, we then analysed local and regional meteorological measurements in order to verify whether critical climatic limits have been exceeded since *Trachycarpus fortunei* has started expanding its range into native vegetation in Europe. For the regional analysis, STASH was applied with the bioclimatic limits derived for the species' native range, using gridded climate data averaged over two different time periods (1931-60 and 1991-2000) for Europe (MITCHELL et al. 2004; 0.16° resolution). The first period was before the enhanced spread of palms and the second is well within the period of establishment and naturalisation of *Trachycarpus fortunei* in southern Switzerland. Thus, we are able to verify, whether the invasion history documented with field reports in southern Switzerland is also reflected in the temporal pattern of changes in climatic conditions (cf. DUKES & MOONEY 1999, WALTHER 2000), or simply was a chorological phenomenon, depending on the availability of seed sources and suitable habitats. The improved knowledge of the history, chronology and driving mechanisms of the observed local establishment of *Trachycarpus* populations south of the Alps was then used as a basis for a better understanding and interpretation of recent occurrences and shifts of *Trachycarpus* and other palm species at their poleward fringe area on the global scale.

4.4 Results

4.4.1 Bioclimatic preferences in the native habitat

Two important bioclimatic factors exclude *Trachycarpus fortunei* in China from higher latitudes and altitudes (Fig. 4.2). Our results show that a mean temperature of + 2.2°C is the threshold for the coldest month in areas with values of about 2300 GDD_5. Assuming a global, generally applicable relationship between monthly mean temperatures and daily extremes (PRENTICE et al. 1992), this corresponds to a minimum temperature of around -19°C. In areas with values of significantly more than 3000 GDD_5, the threshold temperature of the coldest month may be lowered by about half a degree (Fig. 4.2), which suggests a compensatory effect of unfavourable winter temperatures by optimal growth in the growing season.

In accordance with the analysis above, the STASH simulations achieved the best match with the observed distribution, using 2.2°C as the lower limit of the monthly mean temperature of the coldest month and $GDD_5 = 2300$, when the southern/lower limit of the species distribution was defined by a maximum mean temperature of the coldest month of 15.5°C and a tolerated drought index (defined by 1-(actual evapotranspiration/ potential evapotranspiration [AET/PET]; for details see SYKES et al. 1996) of 0.26.

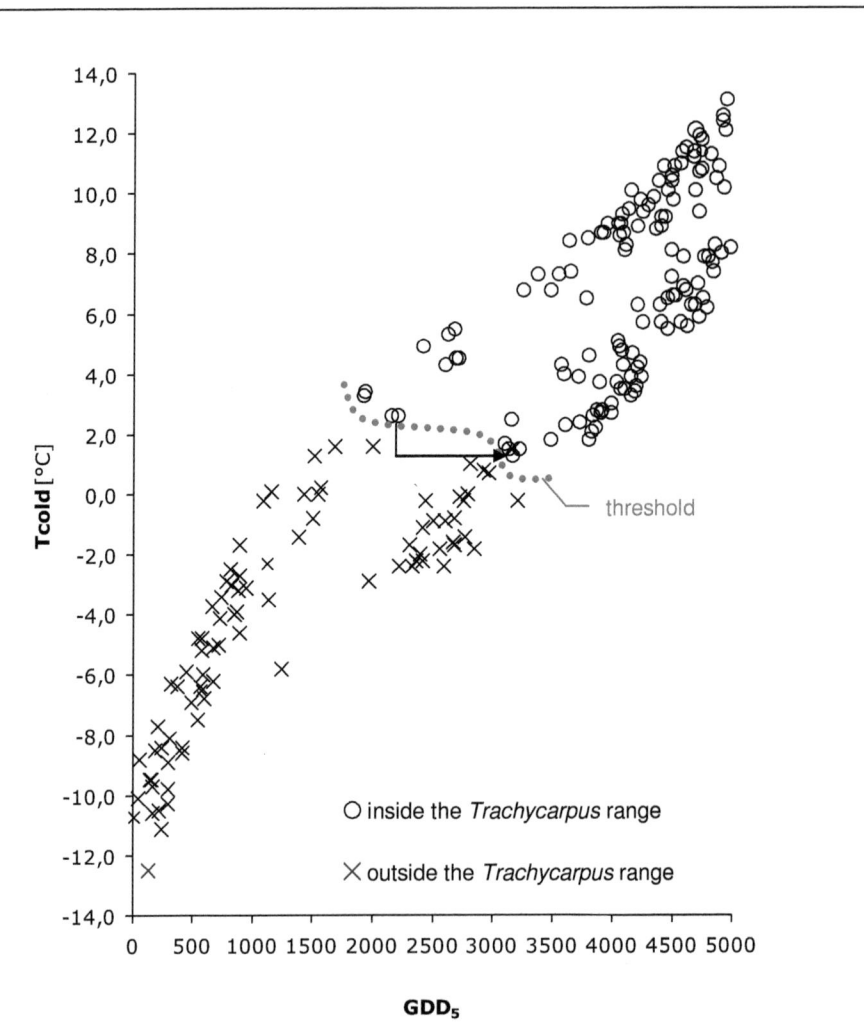

Fig. 4.2: Bioclimatic limits of the distribution of *Trachycarpus fortunei* towards northern and upper range margins in the native habitat in China. (Tcold = monthly mean temperature of the coldest month; GDD_5 = growing degree days above 5°C per year). The arrow indicates a compensatory effect of optimal growth in the growing season for unfavourable winter temperatures (for details see text).

4.4.2 Potential and realised distribution in the introduced range

Numerous local floristic inventories, dating as far back to the 19th nineteenth century, provided data for a detailed reconstruction of the chronology of spread and establishment of a new local palm population south of the Alps (Appendix 2). As in the case of many other ornamental species of the same origin, this palm has been introduced to Europe in the late 18th/early 19th century (JACOBI 1998). However, whereas the introduction and subsequent frequent cultivation in gardens and parks took place about two centuries ago and led to the establishment of large garden populations with fruiting individuals, it is only in the last fifty years that the palm has succeeded in colonising protected sites such as shady and humid gorges. Some twenty years later, first occurrences of palm seedlings in forest stands have been documented, which persisted in the face of competition from the local indigenous flora and reached a fertile stage, so that substantial, fully functioning palm populations were thus established (Fig. 4.3).

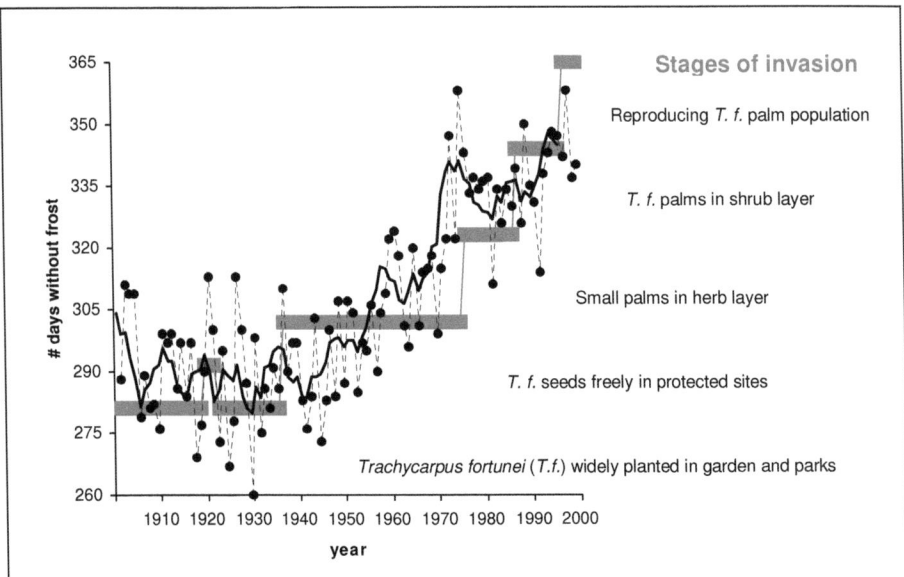

Fig. 4.3: Local climate data vs. invasion history of *Trachycarpus fortunei* palms in southern Switzerland. Milder winters (here indicated by the number of days without frost per year; annual values (broken line & dots) and smoothed 5-years averages (solid line) are shown (data from WALTHER 2002b, modified)) are considered a key factor for the survival and establishment of reproducing palm populations. The later stages of the invasion process have only been reached in the period of warmer climatic conditions (for details see appendix 2 and text).

With the knowledge of the species' specific climatic requirements in the native habitat (Fig. 4.2) and measured climate data of the introduced range, we can now address whether the history of the spatio-temporal spread of *Trachycarpus fortunei* does follow the pattern of improving climatic conditions in the introduced range (Fig. 4.4).

Measured local meteorological data (Fig. 4.4) show that the periods with temperatures above the threshold value of + 2.2°C mean January-temperature (Fig. 4.2) have obviously increased in length and frequency. The smoothed 5-years averages reveal that isolated occasional short-term events with favourable conditions until the 1950s developed into frequent short-term events in the early second half of the 20th century and finally become continuous conditions since the mid 1980s (Fig. 4.4).

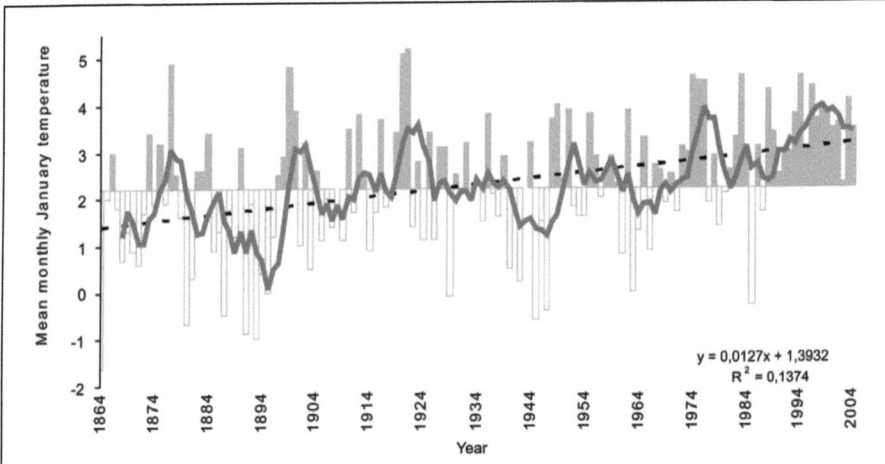

Fig. 4.4: Annual values for mean monthly temperature in January from 1864 to 2004 (data from SMA MeteoSwiss: www.meteoschweiz.ch/web/de/klima/klimaentwicklung/homogene_reihen.html, Meteorological Station of Lugano; BEGERT *et al.* 2005). In addition, the linear trendline (dashed line) as well as smoothed values for five year averages (solid line) are shown.

Furthermore, distinct differences can be seen in the simulated range for *Trachycarpus fortunei* in Europe for the two periods 1931-1960 and for 1991-2000, using the bioclimatic model STASH (Fig. 4.5). There is an obvious shift in the spatial distribution of *Trachycarpus fortunei* in southern-central Europe. In particular, the range of suitable habitats is moving into the area of southern Switzerland during this time (see insets of Fig. 4.5). Whereas in the first period the conditions in southern Switzerland were not suitable to allow enhanced growth and establishment of subspontaneous populations (Fig. 4.5, left), in the latter period the area where *Trachycarpus for-*

tunei has naturalised (southern Ticino/Switzerland) now overlaps with the simulated range of optimal bioclimatic conditions (Fig. 4.5, right).

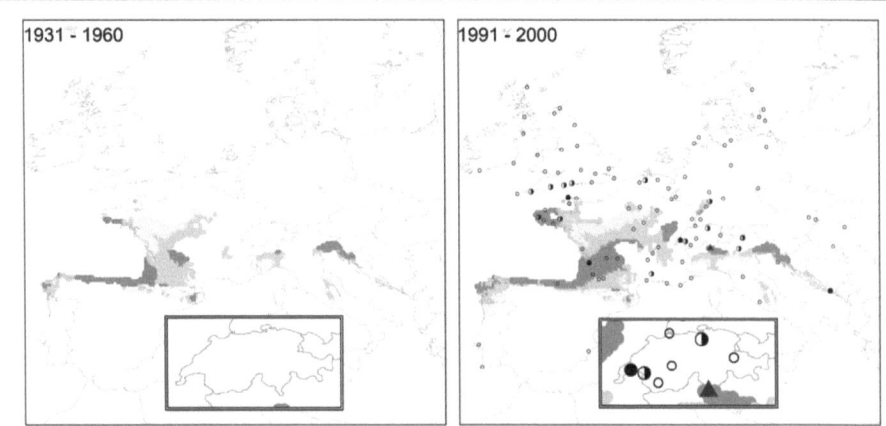

Fig. 4.5: Simulated range for *Trachycarpus fortunei* in Europe for two different periods, based on species-specific bioclimatic limits in the native range. Dark shades of grey denote better conditions for establishment and growth (within the climatic envelope) than light shades due to the direct effect of climate on net assimilation and respiration simulated by the model (SYKES *et al.* 1996). The insets show that in southern Switzerland, the conditions became suitable for *Trachycarpus* palms only in the latter period. For symbols legend see Figure 4.1.

The model output suggests also other areas in Central Europe (e.g. coastal areas at the Bay of Biscay, but also new areas in Central-Western France and at the French-German border) where the conditions from a bioclimatic perspective seem to become increasingly suitable for *Trachycarpus* to grow (cf. also Fig. 4.1).

4.4.3 The new northernmost palm population from a global perspective

With this improved understanding of the history, chronology and driving mechanisms of an observed local establishment of palm populations in the introduced range (Appendix 2; cf. also CARRARO *et al.*, 1999, WALTHER, 2002b, 2003), we are able to analyse and extrapolate these findings to the continental and global perspective.

The new subspontaneous population at the southern foot of the Alps clearly is located outside the known global distribution of palms (Fig. 4.1). This new exclave of palm distribution occurs ca. 300 km NNE beyond the northernmost palm limit as recognised up to now (inset of Fig. 4.1) and its spatio-temporal development strongly suggests a climate change explanation (see above).

Trachycarpus occurrences north of the Alps in Central Europe and in the southern coast of the British Isles can be seen as early stage invasions, where the seeds from planted *Trachycarpus* palms are able to germinate in gardens and parks and to survive at least for some limited time (see Fig. 4.1). On other continents, but still within the belt of global palm distribution, *Trachycarpus fortunei* has been observed to spread out from garden areas into (semi-)natural habitats such as woodlands. Locally well established plants have been reported e.g. from Austin/Texas (L. LOCKETT, pers. comm.) and from North Island of New Zealand (e.g. HEALY & EDGAR 1980) (Fig. 4.1).

4.5 Discussion

It is widely accepted that a distribution including higher latitudes and altitudes of evergreen broad-leaved species in general and palms in particular is limited by the climatic conditions of the cold season (WOODWARD 1987, JONES 1995, WALTHER 2002b, FRANCKO 2003, LÖTSCHERT 2006). For the palm, *Trachycarpus fortunei*, a threshold temperature of about + 2.2°C mean temperature of the coldest month was identified in its native range in China. However, enhanced growing degree accumulations of more than 3000 degree days may compensate unfavourable winter conditions with temperatures down to about + 1.3°C. There is observational evidence from areas outside the native range with suboptimal conditions for the ecological interpretation of such compensatory effects (e.g. WALTHER 2003, FITZROYA 2004). Exposure of planted *Trachycarpus* palms to sub-lethal temperatures results in damage to fronds and spear and/or defoliation. As a consequence, a minimum growing degree day accumulation is required for the palm to recover the damage and resume growth (FITZROYA 2004). If that minimum is not achieved in one growing season, the plant is unable to fully replace damaged tissue, expends stored energy to replace greater lost mass at the cost of growth, and enters a period of decline (FITZROYA 2004). This observational evidence is also supported by experimental work where small palm seedlings were exposed to repeated defoliation (MCPHERSON & WILLIAMS 1998, cf. also ANTEN et al. 2003). Hence, a series of consecutive years with unfavourable climatic conditions will eventually kill smaller plants (cf. WALTHER 2003). On the contrary, limiting frost damage in the winter season reduces the number of GDD_5 necessary to replace mass and resume growth (for further details see FITZROYA 2004).

With this background, it is clear that although the cultivation of adult plants has been possible for decades in the area at the southern foot of the Alps in the past (and more recently in other areas north of the global palm distribution (cf. Fig. 4.1)), the regeneration south of the Alps has not been successful until the last few decades when there were consecutive years of climatic conditions above the critical threshold. This time lag in chronology between introduction and spread, the mechanistic understanding of the ecological impacts of sub-lethal freezing in winter, and the present restriction of sub-spontaneous palm populations to forest stands on southern exposed slopes of lower altitudes (cf. BERGER & WALTHER 2006) strongly suggests that an ameliorated climate, especially warmer and shorter winter seasons, was the essential prerequisite for palms and other evergreen broad-leaved species to become thus locally established (WALTHER 2002b). Species' specific threshold parameters derived from habitat requirements in the native range (see Fig. 4.2) and applied to long-term climate measurements in the introduced range (Fig. 4.4) are in agreement with this climate change explanation (cf. also BEERLING et al. 1995). This explanation is further supported by the results of a bioclimatic model, which highlighted the range of the potential suitable habitat in the introduced range at different periods in the past and present (Fig. 4.5). Periods with a continuous suitable climate have not occurred until the last few decades, allowing species to establish and locally naturalise. Additionally, these new wild populations of palms south of the Alps have reached a life stage and population size that makes them independent of seed supply from planted individuals in gardens and parks, and guarantee long-term survival of a new northernmost palm population, provided the ameliorated climatic conditions of the last few decades continue.

The rejuvenation of *Trachycarpus fortunei* has been observed not only in southern Switzerland, the same species is reported to seed freely in gardens along the southern coast of Great Britain (e.g. C. EVANS, Bournemouth & J. JONES, Truro, pers. comm.), but there, it still is restricted to garden areas and small individuals and, thus, to an early stage of a (potential) invasion process (WALTHER, pers. observ.). In other, more southern areas in Europe, more established *Trachycarpus* populations are reported (e.g. KOVACEVIC 1998), whereas in parts of Australia (e.g. GROVES 1998) and New Zealand (e.g. HEALY & EDGAR 1980, see also NEW ZEALAND PLANT DATABASE (http://nzflora.landcareresearch.co.nz); PETERSON et al. 2006) *Trachycarpus fortunei* is recorded as fully naturalised. In the US, a reproducing *Trachycarpus* population is reported from the southern fringe of Austin (Texas) (L. LOCKETT, pers. comm.). In forest stands in Japan, increases in population sizes of *Trachycarpus fortunei* have recently been reported from as far north as the Tokyo area (KAMEI & OKUTOMI, 1992, FUJIWARA & BOX 1999, KOMURO & KOIKE 2005).

Last but not least, particularly in North America (PARKER 1994, FRANCKO & WILHOITE 2002, see also GILMAN & WATSON 1994) and Europe (STÄHLER 2000, WALTHER 2002a, 2003), but also on other continents – though from the latter less information is publicly available – *Trachycarpus* has increasingly been cultivated even further beyond the potential range of palm distribution (see Fig. 4.1), which makes this particular species a 'cornerstone palm' in many parts of the world (KEMBREY 2004). The expected future global warming may thus not only facilitate the survival of the garden populations (BISGROVE & HADLEY 2002), but – in some areas – these garden populations may act as future dispersal centres for further expansion of palms in response to continued amelioration of climate with global warming, thus allowing the spread and establishment of further palm populations similar to those of the southern foot of the Alps. Outputs from bioclimatic models, which are based on and validated with ground-truth data may help to identify new areas where this process is likely to be detected in the near future, as shown on an European scale in Fig. 4.5.

Not only in Europe with *Trachycarpus*, but world-wide are palm species benefiting from ameliorated climatic conditions. In parallel to the aforementioned situation at the southern foothills of the Alps, palms of other genera are extending elsewhere into new territories north of their former distribution. Palm increases have been noted in south-western North America with the indigenous species *Washingtonia filifera* (CORNETT 1991), with *Sabal mexicana* in Texas (LOCKETT 2004), and with *Sabal minor* in parts of Tennessee and other areas of the South-East where it is not native (D. FRANCKO, pers. comm.) (see also BJORHOLM et al. 2005).

Hence, palms in general, and *Trachycarpus fortunei* in particular, may serve not only as important indicators for the reconstruction of the past climate in Earth history (MAI 1995, cf. also BRÖNNIMANN 2002), they are becoming significant global bioindicators across continents for contemporary climate change and the projected global warming of the near future.

4.6 Acknowledgements

The presented findings have been elaborated in collaboration with DIONEA S.A., environmental consulting office, Locarno/Switzerland, especially G. Carraro & P. Gianoni. Thanks go to S. Shafer, U.S. Geological Survey, Corvallis, Oregon/USA; M. O'Connell, University of Galway/Ireland; M. McGlone, Landcare Research, Lincoln/New Zealand for comments to an earlier version of the manuscript. The correspondence and exchange of expertise with T. Boller, Tauberbischofsheim-Distelhausen/Germany; M. Ferguson, Vancouver/Canada; J. Fitzroya, Colorado Springs, Colorado/USA; D. Francko, Miami University, Oxford, Ohio/USA; M. Gibbons, London/United Kingdom; R. T. Harms, University of Texas, Austin/USA; S. Horne, Greenwood/USA; F. Koike, Yokohama National University, Yokohama/Japan; A. Krüger, Cologne/Germany; L. Lockett, Austin/USA; N. Parker, North Delta, BC/Canada; P. G. Peterson, Landcare Research, Palmerston North/New Zealand; D. Pfenninger, Zurich/Switzerland; M. Turner, Marietta, SC/USA; M. van den Berg, Veenendaal/The Netherlands; P. Vittoz, University of Lausanne, Lausanne/Switzerland; J.-C. Wattenhofer, Chardonne/Switzerland, was very much appreciated and allowed an update of a most present situation of palm populations at the very edge of their distribution. The compilation of several sites with *Trachycarpus fortunei* was made possible only thanks to the manifold reports from palm enthusiasts published in the journals of palm societies and/or in the internet (cf. Appendix 3). Funding by the following agencies is kindly acknowledged: Swiss National Science Foundation (Subproject within NRP 31 & Project Nr. 31-46761.96), German Research Foundation (Project WA 1523/5-1), and by the EC (within the FP 6 Integrated Project "ALARM"; GOCE-CT-2003-506675). The CRU climate data were supplied courtesy of the Climatic Research Unit, University of East Anglia.

The following supplementary material is available for this article:

Appendix 2. Detailed chronology of establishment of the new northernmost palm population (synonyms: *Trachycarpus fortunei* (Hook.) Wendl. = *T. excelsa* Wendl. = *Chamaerops excelsa* Thunb.)

Appendix 3. Localities to the (non-exhaustive) compilation of *Trachycarpus*-sites based upon literature and internet search as well as personal observations and contacts (cf. Fig. 4.1 & 4.5)

Appendix 4. Pictures illustrating the different developmental stages of the new northernmost *Trachycarpus* population

Appendix 5. Realised and modelled (grey shaded areas) distribution of *Trachycarpus fortunei* in its native range, using the following parameters: 2.2°C as the lower limit of the monthly mean temperature of the coldest month and $GDD_5 = 2300$, a maximum mean temperature of the coldest month of 15.5°C and a tolerated drought index of 0.26 (for details see text and Fig. 4.5).

4.7 References

ANTEN, N.P.R., MARTINEZ-RAMOS, M. & D.D. ACKERLY (2003): Compensatory growth in a tropical understorey palm subjected to repeated defoliation events. *Ecology* **84**: 2905-2918.

BEERLING, D.J., HUNTLEY, B. & J.P. BAILEY (1995): Climate and the distribution of *Fallopia japonica*: use of an introduced species to test the predictive capacity of response surfaces. *J. Veg. Sci.* **6**: 269-282.

BEGERT, M., SCHLEGEL, T. & W. KIRCHHOFER (2005): Homogeneous temperature and precipitation series of Switzerland from 1864 to 2000. *Int. J. Climatol.* **25**: 65-80.

BERGER, S. & G.-R.WALTHER (2006): Distribution of evergreen broad-leaved woody species in Insubria in relation to bedrock and precipitation. *Bot. Helv.* **116**: 65-77.

BISGROVE, R. & HADLEY, P. (2002): Gardening in the Global Greenhouse: The Impacts of Climate Change on Gardens in the UK. Technical Report. – UKCIP, Oxford.

BJORHOLM, S., SVENNING, J.-C., SKOV, F. & H. BALSLEV (2005): Environmental and spatial controls of palm (Arecaceae) species richness across the Americas. *Global Ecol. Biogeogr.* **14**: 423-429.

BRÖNNIMANN, S. (2002): Picturing climate change. *Clim. Res.* **22**: 87-95.

CARRARO, G., KLÖTZLI, F., WALTHER, G.-R., GIANONI, P. & R. MOSSI (1999): Observed changes in vegetation in relation to climate warming. Final Report NRP 31. – vdf Hochschulverlag, Zürich.

CORNETT, J.W. (1991): Population dynamics of the palm, *Washingtonia filifera*, and global warming. *San Bernardino County Museum Association Quarterly* **39**: 46-47.

DELECTIS FLORAE REIPUBLICAE POPULARIS SINICAE (1991): Flora Reipublicae Popularis Sinicae 13(1) Palmae. – Science Press, Beijing.

DUKES, J.S. & H.A. MOONEY (1999): Does global change increase the success of biological invaders? *Trends Ecol. Evol.* **14**: 135-139.

EDWARDS, K.R., ADAMS, M.S. & J. KVET (1998): Differences between European native and American invasive populations of *Lythrum salicaria. J. Veg. Sci.* **9**: 267-280.

FITZROYA, J. (2004): Palms in Colorado Springs (USDA Zone 5b). Published online: http://hometown.aol.com/fitzroya/myhomepage/cooking.html

FRANCKO, D.A. (2003): Palms won't grow here and other myths. – Timber Press, Portland.

FRANCKO, D.A. & S.L. WILHOITE (2002): Cold-hardy palms in Southwestern Ohio: Winter damage, mortality and recovery. *Palms* **46**: 5-13.

FUJIWARA, K. & E.O. BOX (1999) Evergreen broad-leaved forests in Japan and eastern North America: Vegetation shift under climatic warming. In: F. KLÖTZLI & G.-R. WALTHER (eds.): Recent shifts in vegetation boundaries of deciduous forests, especially due to general global warming. pp 273-300. – Birkhäuser, Basel.

GIANONI, G., CARRARO, G. & F. KLÖTZLI (1988): Thermophile, an laurophyllen Pflanzenarten reiche Waldgesellschaften im hyperinsubrischen Seenbereich des Tessins. *Ber. Geobot. Inst. ETH, Stiftung Rübel, Zürich* **54**: 164-180.

GIBBONS, M. (2003): A pocket guide to palms. – PRC Publishing Ltd., London.

GILMAN, E.F. & D.G. WATSON (1994): *Trachycarpus fortunei* – Windmill Palm. Fact Sheet ST-645, October 1994. – Southern Group of State Foresters, US Department of Agriculture & Forest Service.

GOOD, R. (1953): The Geography of the flowering plants. 2nd ed. – Longmans, Green and Co, London.

GUISAN, A. & W. THUILLER (2005): Predicting species distribution: offering more than simple habitat models. *Ecology Letters* **8**: 993-1009.

HUGHES, L. (2000) Biological consequences of global warming: is the signal already apparent? *Trends Ecol. Evol.* **15**: 56-61.

JACOBI, K. (1998): Palmen für Haus und Garten. 4th ed. – BLV, München.

JONES, D.L. (1995): Palms throughout the World. – Reed Books, Chatswood.

KAMEI, H. & K. OKUTOMI (1992): Constructive processes of the population of *Trachycarpus fortunei* and its ecological backgrounds in the Institute for Nature Study, Tokyo (I) Characteristics in the distributional expansion of *T. fortunei*. *Reports of the Institute for Nature Study, Tokyo* **23**: 21-36.

KEMBREY, N. (2004): Trachy Troubles. *Chamaerops* **48**: 9-12.

KOMURO, T. & F. KOIKE (2005): Colonization by woody plants in fragmented habitats of a suburban landscape. *Ecological Applications* **15**: 662-673.

KOVACEVIC, M. (1998): The significance of the spontaneous vegetation in the old garden of the arboretum Trsteno (Croatia). *Acta Botanica Croatica* **55/56**: 29-40.

LARCHER, W. & A. WINTER (1981): Frost susceptibility of palms: experimental data and their interpretation. *Principes* **25**: 143-152.

LOCKETT, L. (2004): The Sabal Palm: Restoring a species we didn't know we had (Texas). *Ecological Restoration* **22**: 137-138.

LÖTSCHERT, W. (2006): Palmen: Botanik, Kultur, Nutzung. – Ulmer, Stuttgart.

LÜDI, W. (1949): Bericht über den 6. Kurs in Alpenbotanik. *Ber. Geobot. Inst. ETH, Stiftung Rübel, Zürich* **12**: 12-50.

MAI, D.H. (1995): Tertiäre Vegetationsgeschichte Europas. – G. Fischer, Stuttgart.

MCPHERSON, K. & K. WILLIAMS (1998): The role of carbohydrate reserves in the growth, resilience, and persistence of cabbage palm seedlings (*Sabal palmetto*). *Oecologia* **117**: 460-468.

MITCHELL, T.D., CARTER, T.R., JONES, P.D., HULME, M. & M. NEW (2004): A comprehensive set of high-resolution grids of monthly climate for Europe and the globe: the observed records (1901–2000) and 16 scenarios (2001–2100). *Tyndall Centre Working Paper* 55 – Norwich, UK.

NEW, M., HULME, M. & P.D. JONES (2000): Representing twentieth century space-time climate variability. Part 2: Development of 1901–96 monthly grids of terrestrial surface climate. *J. Clim.* **13**: 2217-2238.

PARKER, N. (1994): Northern limit of palms in North America: *Trachycarpus* in Canada. *Principes* **38**: 105-108.

PARMESAN, C. & G. YOHE (2003): A globally coherent fingerprint of climate change impacts across natural systems. *Nature* **421**: 37-42.

PETERSON, P.G., ROBERTSON, A.W., LLOYD, B. & S. MCQUEEN (2006): Non-native pollen found in short-tailed bat (*Mystacina tuberculata*) guano from the central North Island, New Zealand. *J. Ecol.* **30**: 267-272.

PRENTICE, I.C., CRAMER, W., HARRISON, S.P., LEEMANS, R., MONSERUD, R.A. & A.M. SOLOMON (1992): A global biome model based on plant physiology and dominance, soil properties and climate. *J. Biogeogr.* **19**: 117-134.

ROOT, T.L., PRICE, J.T., HALL, K.R., SCHNEIDER, S.H., ROSENZWEIG, C. & J.A. POUNDS (2003): Fingerprints of global warming on wild animals and plants. *Nature* **421**: 57-60.

SAKAI, A. & W. LARCHER (1987): Frost survival of plants. Ecological Studies **62**. – Springer, Berlin.

SIMBERLOFF, D. (2000): Global climate change and introduced species in United States forests. *The Science of the Total Environment* **262**: 253-261.

SOBRINO VESPERINAS, E., GONZALEZ MORENO, A., SANZ ELORZA, M., DANA SANCHEZ, E., SANCHEZ MATA, D. & R. GAVILAN (2001): The expansion of thermophilic plants in the Iberian Peninsula as a sign of climatic change. In: WALTHER, G.-R. BURGA, C.A. & P.J. EDWARDS (eds.): "Fingerprints" of Climate Change - Adapted behaviour and shifting species ranges. pp. 163-184. – Kluwer Academic/Plenum Publishers, New York.

STÄHLER, M. (2000): Palmen in Mitteleuropa. – The European Palm Society.

SYKES, M.T., PRENTICE, I.C. & W. CRAMER (1996): A bioclimatic model for the potential distributions of north European tree species under present and future climates. *J. Biogeogr.* **23**: 203-233.

WALTHER, G.-R. (2000): Climatic forcing on the dispersal of exotic species. *Phytocoenologia* **30**: 409-430.

WALTHER, G.-R. (2002a): Die Verbreitung der Hanfpalme *Trachycarpus fortunei* im Tessin – 50 Jahre nach der Erstaufnahme. *Schweiz. Beitr. Dendrol* **47**: 29-41.

WALTHER, G.-R. (2002b): Weakening of climatic constraints with global warming and its consequences for evergreen broad-leaved species. *Folia Geobot.* **37**: 129-139.

WALTHER, G.-R. (2003) Wird die Palme in der Schweiz heimisch? *Bot. Helv.* **113**: 159-180.

WALTHER, G.-R. (2004): Plants in a warmer world. *Perspect. Plant Ecol. Evol. Syst.* **6**: 169-185.

WALTHER, G.-R., POST, E., CONVEY, P., MENZEL, A., PARMESAN, C., BEEBEE, T.J.C., FROMENTIN, J.-M., HOEGH-GULDBERG, O. & F. BAIRLEIN (2002): Ecological responses to recent climate change. *Nature* **416**: 389-395.

WINTER, A. (1976): Die Temperaturresistenz von *Trachycarpus fortunei* Wendl. und anderen Palmen. – Dissertation, Universität Innsbruck.

WU, Z.-Y. & DING, T.-Y. (1999): Seed plants of China. –Yunnan Science and Technology Press, Kunming.

5 Bioclimatic limits and range shifts of cold-hardy evergreen broad-leaved species at their northern distributional limit in Europe[4]

5.1 Abstract

The few native evergreen broad-leaved species occurring in central Europe have attracted the interest of generations of scientists; thus, the factors limiting their northern distribution have been well studied. For investigation of climate change-driven range shifts, these climate-sensitive species are particularly well suited. We here analyse recent range shifts of some of the cold-hardiest evergreen broad-leaved species, including both native and introduced species in Europe.

Based on updated field data and outputs from bioclimatic models, we show that the milder winter conditions of the last few decades are consistent with the northward expansion of potential ranges and an increase in the number of evergreen broad-leaved species. At the landscape scale, these species indicate a considerable change in the composition and structure of temperate deciduous forests in various parts of Europe.

Keywords: Range shifts, Tertiary flora, climate change, global warming, deciduous forests, forest structure.

5.2 Introduction

The vegetation of central Europe is relatively poor in native evergreen broad-leaved species compared to temperate regions of other continents. However, these few species have attracted the interest of generations of scientists, and thus, the factors limiting their northern distribution have been well studied in the past. Climate, especially temperature and the length of the growing season, has been pointed out as an important factor determining establishment and survival of evergreen broad-leaved species at their northern range margins, in Europe and worldwide (BOX 1981, WALTER & BRECKLE 1999, WOODWARD *et al.* 2004, POTT 2005). The temperature limits of many evergreen broad-leaved species have been investigated in detail with physiological methods (e.g. LARCHER

---

[4] Published in Phytocoenologia.
E. Schweizerbart'sche Verlagsbuchhandlung: http://www.borntraeger-cramer.de
Full reference: BERGER, S., SÖHLKE, G., WALTHER, G.-R. & R. POTT: Bioclimatic limits and range shifts of cold-hardy evergreen broad-leaved species at their northern distributional limit in Europe. *Phytocoenologia* **37**: 523-539.

1954, 1970, 2000, SAKAI 1982, SAKAI & LARCHER 1987), as well as by biogeographical comparison (e.g. IVERSEN 1944, JÄGER 1975).

The fact that even the most cold-hardy evergreen broad-leaved species are sensitive to low winter temperatures makes them suitable as climate indicators. This fact has been applied to reconstruct past climatic fluctuations from geological records (IVERSEN 1944). Recent climate change has left visible 'fingerprints' within different ecosystems all over the world (WALTHER et al. 2001, PARMESAN & YOHE 2003, ROOT et al. 2003), and it is likely that the warming trend will continue or even increase (IPCC 2001). Climatic indicators serving the reconstruction of past climatic conditions may also be used as indicators of recent and near-future climate change. Climate induced range shifts of individual evergreen broad-leaved species have been detected in different parts of Europe (DIERSCHKE 2005, DOBBERTIN et al. 2005, WALTHER et al. 2005). At the regional scale, an assemblage of evergreen broad-leaved species favoured by climate change has also been investigated in detail (e.g. WALTHER 2000). Whereas the aforementioned studies focus on single species only or on a particular habitat, we here analyse how the cold-hardiest evergreen broad-leaved species, a group including both natives and exotics, shift their northern range margins in concert with recent global warming. The potential ranges implied by species' specific bioclimatic limits are compared with updated data on the realised distributions of these species across Europe.

Because the evergreen broad-leaved plant type has been scarcely represented in central European plant communities, changes in the diversity and distribution of evergreen broad-leaved species not only lead to changes in the physiognomy of existing forest communities, but may in the long run also form new assemblages in different parts of Europe.

5.3 Methods

Recent range extensions at the northern distribution boundaries of selected cold-hardy evergreen broad-leaved species were detected by comparing historical distribution maps and records with present distribution data. Historical data were compiled from ADAMOVIĆ (1909), HOLMBOE (1913), LOESENER (1919), ENQUIST (1924), SCHRÖTER (1936), SCHMUCKER (1942), IVERSEN (1944), BROWICZ (1960), FÆGRI (1960), SCHMID (1956), WALTER & STRAKA (1970), HORVÁT et al. (1974), JÄGER (1975), MEUSEL et al. (1978) and others (for details see BERGER (2003), SÖHLKE (2006) and WALTHER et al. (2007)). The data on the current distribution were compiled from numerous national and regional vegetation databases and surveys, email surveys interviewing experts on plant distribution (cf. acknowledgements), recent literature records and our own field observations. Many of the new literature records were verified in the field, especially occurrences forming new distribution boundaries, in order to confirm the correct identification of the species and to assure that the species is still present in the given locations.

The resulting updated distribution maps were superposed on potential distribution areas of the species based on various bioclimatic parameters. These parameters were derived from

(I) biogeographical literature addressing limiting parameters of the respective species,

(II) climatic parameters of the native range of exotic species, and

(III) results of bioclimatic modelling,

emphasising the functional importance of the relevant particular parameters for the species-specific biological traits (see discussion section).

Distribution maps were compiled, analysed and designed using ESRI ArcView 3.3 and ArcPress.

5.3.1 Selected native and introduced evergreen broad-leaved species

Ilex aquifolium is the northernmost evergreen broad-leaved small tree or shrub in Europe, with severe winter frosts limiting its northern distribution. This can be expressed through the mean temperature of the coldest month, which must be above -0.5 °C (IVERSEN 1944). The potential distribution area of *Ilex aquifolium* was modelled with the bioclimatic model STASH (SYKES et al. 1996) for the period 1981-2000 (WALTHER et al. 2005).

The potential distribution of *Laurus nobilis* was modelled by SVENNING & SKOV (2004), based on three key bioclimatic variables, namely growing degree days (GDD), mean temperature of the coldest month and water balance, using a 'bioclimatic envelope' approach.

Prunus laurocerasus is native to the Balkans and the coast of the Black Sea and Caspian Sea, and it is frequently cultivated in Central Europe as an ornamental shrub. Winter temperature and length of the growing season are important climatic parameters limiting the distribution of *Prunus laurocerasus*. The limiting January mean temperature based on the native range was estimated to -1.2°C. The length of the growing season was expressed through number of months per year with mean temperature above 5°. Hence, the potential range of *Prunus laurocerasus* is additionally limited by a growing season length of at least 8 months, and by annual precipitation which must exceed 550 mm/year.

Bioclimatic limits of the palm *Trachycarpus fortunei* were derived from its native range in China (WALTHER *et al.* 2007), suggesting that mean temperature of the coldest month must exceed +2.2°C, annual GDD must exceed 2300 and drought index (SYKES *et al.* 1996, see also PRENTICE *et al.* 1992) does not exceed 0.26.

The potential distributions of *Prunus laurocerasus* and *Trachycarpus fortunei* are based on climate data from the ALARM dataset (MITCHELL *et al.* 2004, REGINSTER *et al.* 2005). The reference period 1991 – 2000 was regarded as representative of current climatic conditions. The 1990s were the warmest decade since the start of the measurements (IPCC 2001) and, hence of special importance in the context of analyses of global warming-related data. In the same period, most of the new occurrences of all the regarded species outside their former ranges were recorded. The potential ranges of all four of these species were superposed, revealing overlapping areas with especially suitable climatic conditions for evergreen broad-leaved species.

In order to assess how range shifts of a set of evergreen broad-leaved species affect the composition and structure of European deciduous forest ecosystems we compared the evergreen broad-leaved species occurring in southern Switzerland and in Great Britain, both areas where a pronounced increase of evergreen broad-leaved species has taken place in the last few decades.

5.4 Results

The current and potential distributions of the selected evergreen broad-leaved species are shown in Fig. 5.1. The realised historical range (Fig. 5.1, top left: hatched area) of *Ilex aquifolium* is shown according to WALTER & STRAKA (1970) and MEUSEL *et al.* (1978). The new occurrences outside the historical range indicate a shift in the distribution of *Ilex aquifolium* towards the north in Norway and northeast in Germany, Denmark (cf. also BAÑUELOS *et al.* 2004) and southern Sweden, where *Ilex aquifolium* expanded into new areas along the southern Swedish coast. Most new occurrences (Fig. 5.1, top left: black dots) were first recorded in the same time span (1981-2000) as the model refers to (WALTHER *et al.* 2005). The new occurrences on the mainland and larger islands are all within the modelled new potential range.

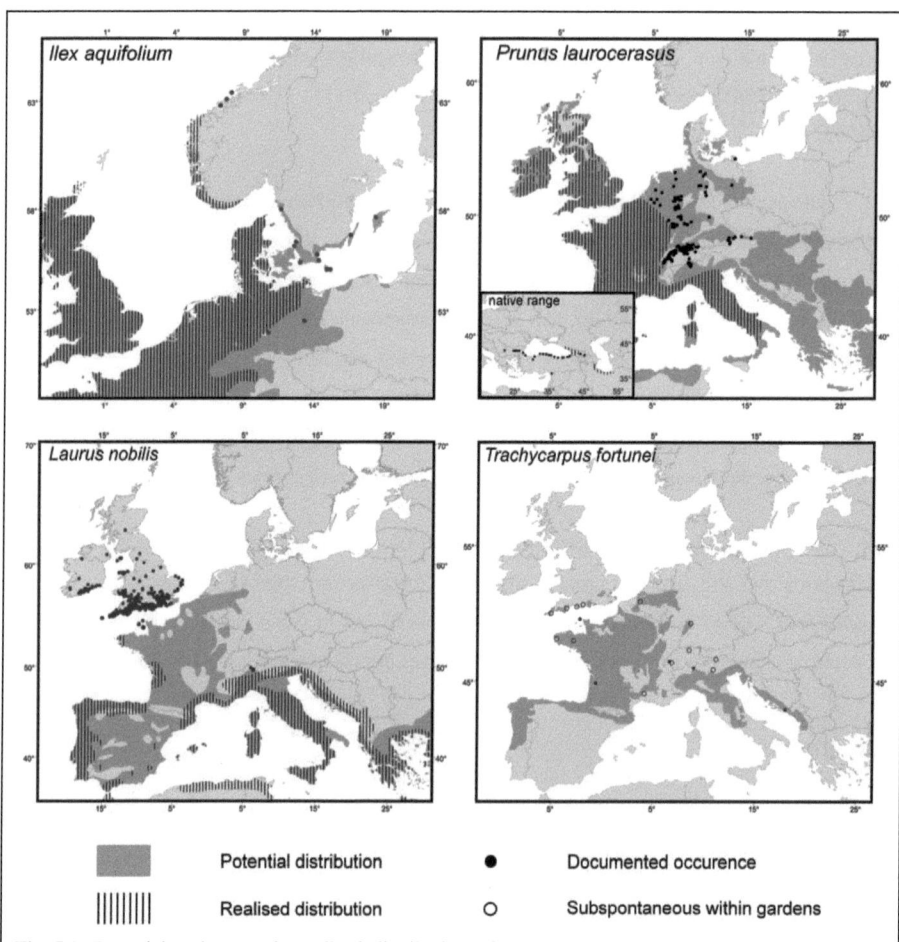

Fig. 5.1: Potential and currently realised distribution of selected evergreen broad-leaved species (for details see text).

The realised distribution of *Prunus laurocerasus* is based on JÄGER (1975) and BSBI (2006) (Fig. 5.1, top right: hatched area); further new occurrences (Fig. 5.1, top right: dots) are compiled from various sources and our own observations (SÖHLKE 2006). These new records are found within large parts of the potential distribution area. The records forming the new northeastern boundary of the introduced range in northern Germany, with single occurrences as far north as Rügen and as far east as Berlin, all originate from the last two decades (SÖHLKE 2006). The number of records from Great Britain also increased substantially in the last decades (see PERRING & WALTERS 1962 and following editions, BSBI 2006).

The realised distribution of *Laurus nobilis* is based on MEUSEL *et al.* (1978), BRULLO *et al.* (2001), BSBI (2006) and WOHLGEMUTH *et al.* (2006); it overlaps with the potential distribution (SVENNING & SKOV 2004) in large areas (Fig. 5.1, bottom left). The historical range of *Laurus nobilis* encompassed the Mediterranean, whereas at present, the species occurs as far north as in Britain and in southern Switzerland. However, to our knowledge there are no confirmed realised occurrences in the northernmost part of the potential distribution area in mainland Europe.

The distribution records of *Trachycarpus fortunei* from Europe are listed in WALTHER (2003), BSBI (2006) and WALTHER *et al.* (2007). Escaped *Trachycarpus fortunei* palms have been recorded in southern Switzerland in large numbers, but also in France, as far north as Brittany, and rejuvenation within gardens has been observed in southern England. Also the northernmost sub-spontaneous occurrences of *Trachycarpus fortunei* correspond with the boundary of the potential distribution area.

In addition to individual species distributions, Fig. 5.2 shows the overlap of the potential distribution areas of all the species shown in Fig. 5.1. The potential species distributions overlap in substantial parts of central Europe, with gradually decreasing number of species' ranges towards the northeast. Whereas updated distribution data from France are scarce, more current information on evergreen broad-leaved species is available from the British Isles and Switzerland. Both the southernmost part of Switzerland and southern Great Britain have especially favourable winter temperatures, as compared with other parts of Europe where deciduous forests occur. This is reflected in the coincidence of potential distribution areas of the four species regarded here, and additional evergreen broad-leaved species, which have also been recorded in these areas. A non-exhaustive list of additional evergreen broad-leaved species recorded in deciduous forests of southern Great Britain and southern Switzerland is given in Table 5.1. Many of the same (exotic) species occur in both areas, but some species are restricted to Great Britain (e.g. *Rhododendron ponticum*) and others to southern Switzerland (e.g. *Cinnamomum glanduliferum, Pittosporum tobira*).

Table 5.1: Evergreen broad-leaved species recorded in deciduous forests of Europe. Records from Switzerland according to WALTHER (1999) and from Great Britain according to BSBI (2006). Frost resistance refers to the experimental work of SAKAI (1982), SAKAI & LARCHER (1987) and DIRR & LINDSTROM (1990).

Species name	Origin	Frost resistance	Switzerland	Great Britain
Ilex aquifolium L.	Central Europe	-23	+	+
Hedera helix L.	Central Europe	-15	+	+
Quercus ilex L.	Mediterranean	-15	+	+
Laurus nobilis L.	Mediterranean	-10	+	+
Viburnum tinus L.	Mediterranean	-10	+	+
Rhododendron ponticum L.	Asia Minor			+
Pyracantha coccinea ROEM.	Mediterranean / Asia Minor		+	+
Prunus laurocerasus L.	Asia Minor	-24	+	+
Mahonia aquifolium (PURSCH) NUTT.	North America		+	+
Aucuba japonica THUNB.	Southeast Asia	-20	occasional	+
Ligustrum lucidum AIT.	Southeast Asia	-15	+	+
Elaeagnus pungens THUNB.	Southeast Asia	-15	+	occasional
Lonicera spp.	Southeast Asia		+	+
Cotoneaster salicifolius FRANCH.	Southeast Asia		+	+
Viburnum rhytidophyllum HEMSL.	Southeast Asia		occasional	+
Trachycarpus fortunei (HOOK) WENDL.	Southeast Asia	-14	+	only within gardens, except on the island of Alderney
Euonymus japonica THUNB.	Southeast Asia	-12	occasional	+
Cinnamomum glanduliferum (WALLICH.) MEISSN.	Southeast Asia	-13	+	
Eriobotrya japonica (THUNB.) LINDL.	Southeast Asia	-10	occasional	occasional
Pittosporum tobira AIT.	South east Asia	-10	+	

In both Britain and southern Switzerland, the number, developmental stage and abundance of evergreen broad-leaved species varies with latitude and altitude. Relevés from WALTHER (2000) and BERGER & WALTHER (2006) reveal an altitudinal gradient in the number and abundance of evergreen broad-leaved species on south-facing slopes in the Insubrian region (Fig. 5.3). At low altitudes there are dense stands of evergreen broad-leaved species in the shrub layer, and some individuals of e.g. *Laurus nobilis* and *Cinnamomum glanduliferum* have grown up to the tree layer. Towards higher altitudes the evergreen broad-leaved species are restricted to the shrub layer and their cover decreases. Finally there are only single small specimens in the herb layer, and at even higher altitudes no evergreen broad-leaved species are growing at all. In parallel, the number of evergreen broad-leaved species present declines with altitude.

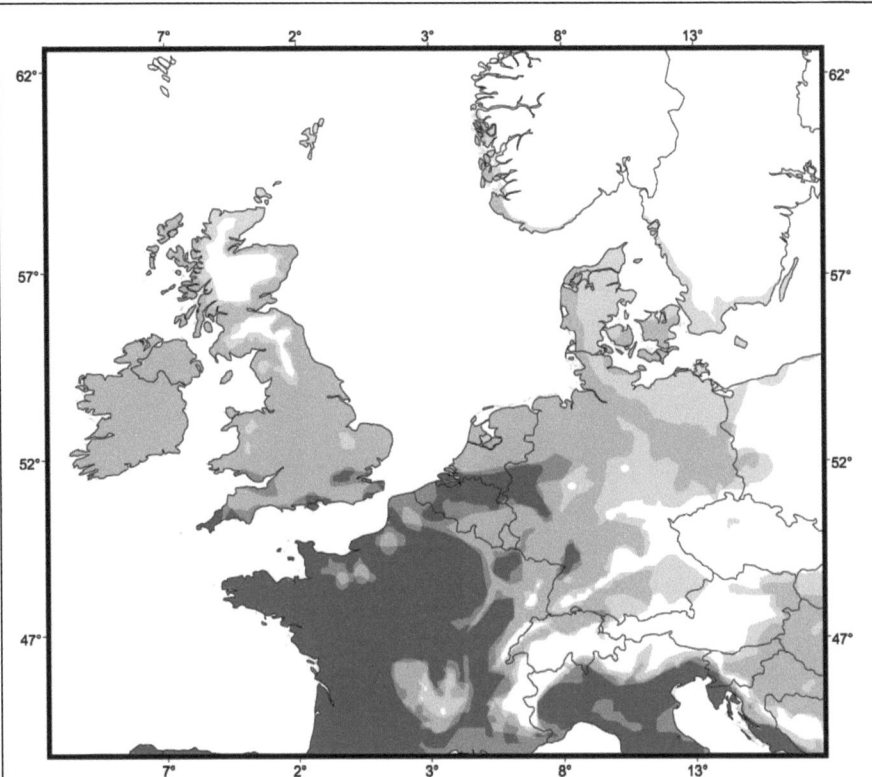

Fig. 5.2: Superposed potential distribution of the four evergreen broad-leaved species shown in Fig. 5.1. Shading represents the number of evergreen broad-leaved species (Overlap of all four species' potential ranges indicated by the darkest shading, decreasing gradually with decreasing number of species' potential ranges overlapping).

5.5 Discussion

Previous studies showed that climate change has favoured the spread of evergreen broad-leaved species in different parts of Europe. For instance, in southern Switzerland and northern Italy an increasing number of evergreen broad-leaved species was recorded as a consequence of climate change (GIANONI et al. 1988, WALTHER 2000, 2001, 2002). Also, the northward range shift of *Ilex aquifolium* in Scandinavia (WALTHER et al. 2005) and the altitudinal upward shift of *Viscum album* in the European Alps (DOBBERTIN et al. 2005) are regarded as consequences of warmer winter temperatures. In Central European deciduous forests, the change of the growth form of *Hedera helix* from creeping along the ground to climbing up trees has likewise been linked to recent climate change (DIERSCHKE 2005).

We here regard this phenomenon more broadly, for a group of species, and present a biogeographic analysis that shows an emerging assemblage of advancing evergreen broad-leaved species over much of Central Europe. The specific bioclimatic limits of these species and the impacts of this ongoing process at landscape scale are discussed.

5.5.1 European distribution and bioclimatic limits

Winter temperature plays an important role in limiting the distribution of the evergreen broad-leaved vegetation towards the poles (RÜBEL 1930, SCHIMPER 1935, SCHMITHÜSEN 1976, BOX 1981, SAKAI & LARCHER 1987, WOODWARD 1987, WALTER & BRECKLE 1999). Although minimum temperature of -15°C is regarded to be an important threshold value for cold-hardy evergreen broad-leaved species in general, the specific climatic threshold differs with the particular biological traits of the given species (e.g. SAKAI & LARCHER 1987, WALTHER 1999). In addition to low winter temperature, other climatic parameters may also limit the species' persistence at northern range margins, such as the length of the growing season (as required to produce viable seeds) and total summer warmth (indicated by GDD; as needed for a positive annual carbon balance and accumulation of biomass).

The most obvious effect of rising winter temperatures is a reduced risk of lethal frost incidents, which allows survival beyond the former range boundary. Additionally, the frequency of sub-lethal frosts, that nevertheless may be severe enough to damage tissue, decreases. Hence, sub-lethal damage, which must be compensated by replacement of damaged tissue, and thus cost energy, will decrease. This favours species at their range margin and thus, survival in formerly unsuitable areas may become possible under climatic warming. Evergreen broad-leaved species are particularly favoured by milder winter temperatures, as they are able to profit from positive net photosynthesis in periods with favourable climatic conditions even in winter (ZELLER 1951, FISCHER & FELLER 1994, OLIVEIRA & PEÑUELAS 2004), in contrast to deciduous trees which are leafless in the winter season.

Although all the four species presented in Fig. 5.1 are restricted by cold winters (with frost resistance varying considerably between species), their temperature limits show that other climatic parameters may also play a role in limiting their distributions, due to specific biological traits. Whereas the survival and regeneration of *Ilex aquifolium* is mainly restricted by severe winters, *Prunus laurocerasus* additionally needs a sufficiently longer growing season in order to produce viable seeds (ADAMOVIĆ 1909, SÖHLKE 2006). Hence, reproduction of *Prunus laurocerasus* is limited by the length of the growing season, which restricts it from extending as far north as *Ilex aquifolium*, in spite of its ability to survive equally low winter temperatures.

The northward extension of *Laurus nobilis* is also restricted by cold winters, though the limiting winter temperature used in the model to project the potential range of *Laurus nobilis* may be overestimated (SVENNING, pers. comm.). The importance of winter temperatures for the northern distributional boundary is supported by the findings of SAKAI & LARCHER (1987), who regarded *Laurus nobilis* as a Mediterranean broad-leaved tree with medium frost resistance, especially pointing out the limited frost resistance of overwintering propagative organs.

Low winter temperatures limit the distribution of *Trachycarpus fortunei* as well. Furthermore, its cold hardiness also depends on sufficiently warm summers (here expressed as GDD). Warm summers allow this palm to accumulate enough biomass to regenerate damaged leaves and thereby to compensate for sub-lethal frost damage of the winter season (WALTHER et al. 2007). The northernmost sub-spontaneous occurrences of *Trachycarpus fortunei* coincide well with the northern boundary of the predicted potential distribution area. The lack of naturalised occurrences of *Trachycarpus fortunei* in southernmost Europe, even though the species is cultivated in this area, reflects the summer dryness of the Mediterranean climate with drought too severe for growth of this species (c.f. BERGER & WALTHER 2006).

At local scale, not only macroclimatic limits but also other ecologically relevant parameters, e.g. edaphic factors, must be taken into account, as they also influence the diversity and structure of communities with an increasing share of evergreen broad-leaved species (BERGER & WALTHER 2006). This is of special importance at species range margins, as the habitat requirements of species become more specific towards range boundaries (THOMAS et al. 2001). Towards the Mediterranean the amount and the temporal pattern of precipitation also plays a role limiting the distribution of evergreen broad-leaved species (BERGER & WALTHER 2006, DEL RIO & PENAS 2006).

5.5.2 Impacts at the landscape level

Vegetation modelling has projected range shifts both by single species and by biomes (e.g. PRENTICE et al. 1992, BAKKENES et al. 2002, WOODWARD et al. 2004, THUILLER et al. 2005, OHLEMÜLLER et al. 2006). The fact that responses to climate change are species-specific, as results presented here show, suggests that climate change in the longer term will lead to a reorganisation of existing communities rather than to synchronous shifts of whole vegetation units. Furthermore, as there are only few native evergreen broad-leaved species in northern and central-European plant communities, the aforementioned range shifts may lead to distinct changes in the diversity and physiognomy of existing forest communities, or to assemblages of new species combinations (CHAPIN et al. 1993, SCOLES & VAN BREEMEN 1997). The substantial overlap of the projected ranges of several evergreen broad-leaved species in southern Great Britain and southern Switzerland (shown in Fig. 5.2) suggests a trend towards more evergreen broad-leaved species in deciduous forests, which is further substantiated by the occurrence of additional evergreen broad-leaved species recorded in these areas (see Table 5.1). A shift from deciduous forest to deciduous forest with more evergreen species may change some forest ecosystem processes, e.g. carbon cycling and water balance.

A new type of deciduous forest, with evergreen understorey, has been described by DELARZE et al. (1999) for southern Switzerland. A structural scheme for this forest type with an increasing number of evergreen broad-leaved species is shown in Fig. 5.3. As species approach their bioclimatic limits towards higher altitudes the size of the individuals and the number of species decreases. Analogously, the number of evergreen broad-leaved species declines towards higher latitudes. Also the growth form of the most cold-hardy species changes towards their northern range margins. For example, *Ilex aquifolium* forms trees in the core area of its distribution but grows more often as a shrub towards the range margin (BERGER 2003). The change in growth form of *Ilex aquifolium* near its range boundary shows that the species is able to grow and survive in a transitional area, but does not reach the size of the individuals in the core area. Similar observations have been registered for *Hedera helix* (see DIERSCHKE 2005) as well as for *Prunus laurocerasus* in its native range, where it occurs only as vegetatively spreading shrubbery at high altitudes (NAKHUTSRISHVILI 1999).

On the British Isles there is no pronounced altitudinal gradient, but the evergreen broad-leaved species may constitute a considerable proportion of the shrub and lower tree layer, e.g. in both southwestern England and Ireland, as has been observed in southern Switzerland. Small evergreen forests and scrub stands (often with *Quercus ilex*) have been described as recently generated vegetation types with no counterpart in the native vegetation, whereas *Rhododendron* thickets and understoreys are matched in the native vegetation by evergreen thickets of *Ilex aquifolium* (PETERKEN 2001).

The overall assessment of the impacts of spreading evergreen broad-leaved species in deciduous forest communities should consider existing communities formed by the respective species in their native ranges as well as those reported in the palaeoecological literature. In North America and Asia species-rich mixed broad-leaved forests are a general phenomenon in latitudes with analogous conditions (FUJIWARA & BOX 1994). In Europe, however, repeated glaciations have reduced the formerly diverse Tertiary evergreen flora to a few species able to recolonise central Europe during the Holocene (SCHMID 1939, LANG 1994). The assemblage of currently spreading evergreen broad-leaved species might be divided into three subgroups based on their origin: (I) species native to Europe, expanding their previous ranges by themselves but also facilitated by human introductions, e.g. *Ilex aquifolium* and *Laurus nobilis*; (II) species widespread throughout the Tertiary not able to recolonise Central Europe in the Holocene, but reintroduced as ornamentals from nearby refugial areas, e.g. *Prunus laurocerasus*; and (III) species introduced from other more distant parts of the world, e.g. *Trachycarpus fortunei, Cinnamomum glanduliferum*.

The now expanding evergreen broad-leaved species represent plant types as well as families characteristic of the European Tertiary flora. The laurel type (Daphno-Macrophanerophyta) was the most important tree form in Tertiary Europe (MAI 1995). Most species of this type are shade tolerant (MAI 1995) and hence may thrive in the understorey of deciduous forests. This still applies to *Ilex aquifolium*, which grows in the understorey of deciduous forests in large parts of its distribution area. In north-western Europe *Ilex aquifolium* grows in different communities of the Querco-Fagetea, also in Atlantic beech and mixed forests like Abieti-Fagetum and Galio-Abietum, and in montane beech forests of Italy (Fagion australo-italicum) (POTT 1990). In south-western Europe beech forests (Rusco-Fagetum, Ilici-Fagetum) have different laurophyllous and sclerophyllous species in the understorey (e.g. *Ruscus aculeatus, Hedera helix, Buxus sempervirens* and *Ilex aquifolium*) (DURIN et al. 1967, TOMBAL 1972, RAMEAU 1985, GÉHU & JULVE 1989). Finally, *Ilex aquifolium* also grows in the understorey of deciduous forests of the Balkan Peninsula, together with *Prunus laurocerasus, Hedera helix* and *Rhododendron ponticum*, where *Fagus orientalis* and *Fagus moesiaca* form the tree layer (Rhododendro pontici-Fagetum orientalis) (HORVÁT et al.

1974). *Prunus laurocerasus* is associated as an evergreen shrub to the understorey of deciduous beech forests (e.g. Fageta laurocerasosa, Fageta luzulosa, Fageto-Abieteta laurocerasosa) in its native range (NAKHUTSRISHVILI 1999).

Numerous fossil leaves, flowers and fruits from several *Cinnamomum* species have been recorded from the Tertiary, showing that the genus *Cinnamomum* (Lauraceae) played an important role in the Tertiary flora of Europe (STAUB 1905, c.f. also KOVAR-EDER *et al.* 2006). All *Cinnamomum* tree species require ample precipitation, in addition to warm temperatures, which the introduced *Cinnamomum glanduliferum* enjoys in the Insubrian region under present conditions (BERGER & WALTHER 2006). Today the genus is restricted to warm-temperate and tropical east and southeast Asia (monsoon area) and to Australia (two species) (STAUB 1905), though some species have become naturalised elsewhere (e.g. FLEISCHMANN 1997, WARD & LABISKY 2004).

Another life form of importance in Europe throughout the Tertiary was the tropical tuft-tree type Macrophanerophyta scaposa ("Schopfbaum"), a single-stemmed tree with a terminal rosette of leaves that die and extend the trunk, as in palms (BOX 1981, MAI 1995). In the recent European flora there are few examples of this type, e.g. *Chamaerops, Phoenix* and *Dracaena*, but the introduced naturalising species *Trachycarpus fortunei* also belongs to this type (cf. Fig. 5.4).

Fig. 5.4a illustrates some of the main aspects of the Tertiary flora of central Europe (HEER 1946). The large tree on the left is *Cinnamomum* and next to it is a small *Laurus* (shaded darker). Also palms like *Sabal* and *Phoenicites* are represented. Fig. 5.4b shows a deciduous forest stand in southern Switzerland in winter, also with a *Cinnamomum* tree, *Laurus* and a palm species (*Trachycarpus fortunei*).

Fig. 5.4: a) Tertiary flora "Lausanne zur miocaenen Zeit" (HEER 1946): Large tree on the left: *Cinnamomum*, small tree on the left: *Laurus* (darker shaded). Palms, e.g. *Sabal* (fan palm), *Phoenicites* (feather palm). b) Deciduous forest stand with evergreen understorey in southern Switzerland (winter), also with *Cinnamomum, Laurus* and the fan palm *Trachycarpus fortunei*.

Recent climate change has favoured the evergreen broad-leaved plant functional type within deciduous forests. With continued warming this process is likely to proceed and to induce changes in the composition and structure of temperate deciduous forests in various parts of Central Europe. Other human activities also cause changes to the native vegetation, for example the introduction of new species. The combination of these anthropogenic influences has facilitated a vegetation type that may symbolise a revival of elements of the Tertiary flora in Europe.

5.6 Acknowledgements

The contribution of distribution data and local expertise from numerous persons enabled this study. Floristic data were kindly provided by T. Tyler (Projekt Skånes Flora, Sweden), L. Jonsell (Projekt Upplands Flora, Sweden), S. Svensson (Gotland, Sweden), G. Weimarck (Göteborg, Sweden), C.-A. Hæggström (Department of Ecology, University of Helsinki, Finland), P.H. Salvesen

(The Norwegian Arboretum, University of Bergen), A. Skogen (University of Bergen, Norway), J. Kollmann and J.M. Bañuelos (Department of Ecology, Royal Veterinary and Agricultural University of Denmark, Copenhagen), G. Stachowiak (Salzwedel, Germany), D. Frank (Halle, Germany), R. May, F. Klingenstein and H. Weber (German Federal Agency for Nature Conservation, Bonn), W. Härdtle (Institute of Ecology and Environmental Chemistry, University of Lüneburg, Germany), M. Diekmann (Vegetation Ecology and Conservation Biology, University of Bremen, Germany), P. Schönfelder and M. Scheuerer (ZdfK, Germany), H.-H. Poppendieck and J. v. Prondzinski (Hamburg, Germany), F. Schacherer (NLWKN, Hannover, Germany), T. Breunig (Karlsruhe, Germany). F.-J. Weicherding (Heiligenwald/Saar, Germany), A. Hoppe and F. Fiebelkorn (Hannover, Germany), F. Willemsen (Elmpter Schwalmbruch, Germany), M. Sonnberger (Heiligkreuzsteinach, Germany), H. Leschus (Wuppertal, Germany), K. Adolphi (Köln, Germany), C. Leitz (Eberbach, Germany), R. Söhlke (Emden, Germany), G. Schmitz (Konstanz, Germany), A. Gerlach (Clausthal-Zellerfeld, Germany), I. Hetzel (Recklinghausen, Germany), B. Bäumler (ZDSF, Switzerland), G. Carraro (DIONEA SA, Locarno, Switzerland), E. Landolt (Institute of Integrative Biology, ETH Zürich, Switzerland), B. Liebst Reber (Kriens, Switzerland), P. Vittoz (Department of Ecology and Evolution, University of Lausanne, Switzerland), R. Zäch (Zürich, Switzerland), F. Essl (Umweltbundesamt, Wien, Austria), M. Malicky (ZOBODAT, Linz, Austria), L. Depypere (Gent, Belgium) F. Verloove (National Botanic Garden, Belgium) and W. van der Slikke (FLORON, Leiden, Netherlands). We also thank E.O. Box and H. Dierschke for their comments to an earlier version of the manuscript.

Funding by the following agencies is kindly acknowledged: German Research Foundation (Project WA 1523/5-1), German Federal Agency of Nature Conservation (FKZ 80581001) and the EC (within the FP 6 Integrated Project 'ALARM'; GOCE-CT-2003-506675).

5.7 References

ADAMOVIĆ, L. (1909): Die Vegetationsverhältnisse der Balkanländer. – Engelmann, Leipzig.

BAKKENES, M., ALKEMADE, J.R.M., IHLE, F., LEEMANS, R. & J.B. LATOUR (2002): Assessing effects of forecasted climate change on the diversity and distribution of European higher plants for 2050. *Glob. Chang. Biol.* **8**: 390-407.

BAÑUELOS, M.J., KOLLMANN, J., HARTVIG, P. & M. QUEVEDO (2004): Modelling the distribution of *Ilex aquifolium* at the north-eastern edge of its geographical range. *Nord. J. Bot.* **23**: 129-142.

BERGER, S. (2003): *Ilex aquifolium* – Bioindikator für Klimaveränderung? – Diplomarbeit, Institut für Geobotanik, Universität Hannover.

BERGER, S. & G.-R. WALTHER (2006): Distribution of evergreen broad-leaved woody species in Insubria in relation to bedrock and precipitation. *Bot. Helv.* **116**: 65-77.

BOX, E.O. (1981): Macroclimate and plant forms: An introduction to predictive modelling in phytogeography. Tasks for Vegetation Science 1. – Junk Publishers, The Hague.

BROWICZ, K. (1960): Immergrüne Laubgehölze in Bulgarien. *Rhododendronges. Bremen, Jahrbuch* **1960**: 18-26.

BRULLO, S., COSTANZO, E. & V. TOMASELLI (2001): Phytosociological study on the *Laurus nobilis* communities in the Hyblaean Mountains (SE Sicily). *Phytocoenologia* **31**: 249-270.

BSBI (2006): BSBI Atlas Update Project. – Botanical Society of the British Isles. URL: http://www.bsbiatlas.org.uk/main.php.

CHAPIN, F.I., RINCON, E. & P. HUANTE (1993): Environmental responses of plants and ecosystems as predictors of the impact of global change. *J. Biosciences* **18**: 515-524.

DELARZE, R., GONSETH, Y. & P. GALLAND (1999): Lebensräume der Schweiz. – Ott Verlag, Thun.

DEL RIO, S. & A. PENAS (2006): Potential areas of evergreen forests in Castile and Leon (Spain) according to future climate change. *Phytocoenologia* **36**: 45-66.

DIERSCHKE, H. (2005): Laurophyllisation - auch eine Erscheinung im nördlichen Mitteleuropa? Zur aktuellen Ausbreitung von *Hedera helix* in sommergrünen Laubwäldern. *Ber. Reinh.-Tüxen-Ges.* **17**: 151-168.

DIRR, M.A. & O.M. LINDSTROM (1990): Leaf and stem cold hardiness of 17 broadleaf evergreen taxa. *J. Environ. Horticult.* **8**: 71-73.

DOBBERTIN, M., HILKER, N., REBETEZ, M., ZIMMERMANN, N.E., WOHLGEMUTH, T. & A. RIGLING (2005): The upward shift in altitude of pine mistletoe (*Viscum album* ssp. *austriacum*) in Switzerland – the result of climate warming? *Int. J. Biometeorol.* **50**: 40-47.

DURIN, L., GÉHU, J.M., NOIRFALISE, A. & N. SOUGNEZ (1967): Les hêtraies atlantiques et leur essaim climatique dans les nordouest de la France. *Bull. Soc. Bot. Nord de la France* **20**: 59-89.

ENQUIST, F. (1924): Sambandet mellan klimat och växtgränser. *Geol. Fören. Förhandl.* **46**: 202-213.

FÆGRI, K. (1960): Maps of distribution of Norwegian plants. – I. The coast plants. – Oslo University Press, Oslo.

FISCHER, A. & U. FELLER (1994): Seasonal changes in the pattern of assimilatory enzymes and proteolytic activities in leaves of juvenile ivy. *Ann. Bot.* **74**: 389-396.

FLEISCHMANN, K. (1997): Invasion of alien woody plants on the islands of Mahe and Silhouette, Seychelles. *J. Veg. Sci.* **8**: 5-12.

FUJIWARA, K. & E.O. BOX (1994): Evergreen broad-leaved forests of the Southeastern United States. – In: MIYAWAKI, A., IWATSUKI, K. & M.M.GRANDTNER (eds.): Vegetation in Eastern North America. pp. 273-312. – University of Tokyo Press, Tokyo.

GÉHU, J.M. & P. JULVE (1989): Die atlantischen Wälder mit Buche: Struktur, Pflanzengeographie, Ökologie, Dynamik und Syntaxonomie. *Ber. Reinh.-Tüxen-Ges.* **1**: 93-106.

GIANONI, G., CARRARO, G. & F. KLÖTZLI (1988): Thermophile, an laurophyllen Pflanzenarten reiche Waldgesellschaften im hyperinsubrischen Seenbereich des Tessins. *Ber. Geobot. Inst. ETH, Stiftung Rübel, Zürich* **54**: 164-180.

HEER, O. (1946): Die Urwelt der Schweiz. – Francke, Bern.

HOLMBOE, J. (1913): Kristtornen i Norge. En plantegeografisk undersøkelse. *Bergens Museums Aarbok* **7**: 1-91.

HORVÁT, J., GLAVAČ, V. & H. ELLENBERG (1974): Vegetation Südeuropas. – Fischer, Jena.

IPCC (2001): Climate Change 2001: The scientific basis. Contribution of working group I to the third assessment report of the intergovernmental panel on climate change. – Cambridge University Press, Cambridge.

IVERSEN, J. (1944): *Viscum, Hedera* and *Ilex* as climate indicators. *Geol. Fören. Förhandl.* **66**: 463-483.

JÄGER, E. (1975): Wo liegen die Grenzen der Kulturareale von Pflanzen. *Wiss. Beitr. Martin Luther-Universität Halle-Wittenberg* **6** (P4): 101-113.

KOVAR-EDER, J., KVAČEK, Z., MARTINETTO, E. & P. ROIRON (2006): Late Miocene to Early Pliocene vegetation of southern Europe (7-4 Ma) as reflected in the megafossil plant record. *Palaeogeogr. Palaeoclimatol. Palaeoecol.* **238**: 321-339.

LANG, G. (1994): Quartäre Vegetationsgeschichte Europas. – Gustav Fischer, Jena.

LARCHER, W. (1954): Die Kälteresistenz Mediterraner Immergrüner und ihre Beeinflussbarkeit. *Planta* **44**: 607-635.

LARCHER, W. (1970): Kälteresistenz und Überwinterungsvermögen Mediterraner Holzpflanzen. *Oecol. Plant.* **5**: 267-285.

LARCHER, W. (2000): Temperature stress and survival ability of Mediterranean sclerophyllous plants. *Plant Biosystems* **134**: 279-295.

LOESENER, T. (1919): Über die Aquifoliaceen, besonders über *Ilex*. *Mitt. Dtsch. Dendrol. Ges.* **28**: 1-69.

MAI, D.H. (1995): Tertiäre Vegetationsgeschichte Europas. – Fischer, Jena.

MEUSEL, H., JÄGER, E., RAUSCHERT, S. & E. WEINERT (1978): Vergleichende Chorologie der Zentraleuropäischen Flora. – Fischer, Jena.

MITCHELL, T.D., CARTER, T.R., JONES, P.D., HULME, M. & M. NEW (2004): A comprehensive set of high-resolution grids of monthly climate for Europe and the globe: the observed record (1901-2000) and 16 scenarios (2001-2100). *Tyndall Centre Working Paper* **55**. – Tyndall Centre for Climate Change Research, University of East Anglia, Norwich, UK.

NAKHUTSRISHVILI, G. (1999): The vegetation of Georgia (Caucasus). *Braun-Blanquetia* **15**: 1-74.

OHLEMÜLLER, R., GRITTI, E.S., SYKES, M.T. & C.D. THOMAS (2006): Quantifying components of risk for European woody species under climate change. *Glob. Chang. Biol.* **12**: 1788-1799.

OLIVEIRA, G. & J. PEÑUELAS (2004): Effects of winter cold stress on photosynthesis and photochemical efficiency of PSII of the Mediterranean *Cistus albidus* L. and *Quercus ilex* L. *Plant Ecol.* **175**: 179-191.

PARMESAN, C. & G. YOHE (2003): A globally coherent fingerprint of climate change impacts across natural systems. *Nature* **421**: 37-42.

PERRING, F.H. & S.M. WALTERS (1962): Atlas of the British Flora. – Thomas Nelson and Sons, London.

PETERKEN, G.F. (2001): Ecological effects of introduced tree species in Britain. *For. Ecol. Manage.* **141**: 31-42.

POTT, R. (1990): Die nacheiszeitliche Ausbreitung und heutige pflanzensoziologische Stellung von *Ilex aquifolium* L. *Tuexenia* **10**: 497-512.

POTT, R. (2005): Allgemeine Geobotanik. – Springer, Berlin.

PRENTICE, I.C., CRAMER, W., HARRISON, S. P., LEEMANS, R., MONSERUD, R.A. & A.M. SOLOMON (1992): A global biome model based on plant physiology and dominance, soil properties and climate. *J. Biogeogr.* **19**: 117-134.

RAMEAU, J.C. (1985): L'intérêt chorologique des quelques groupements forestiers du Morvan, France. *Vegetatio* **59**: 47-65.

REGINSTER, I., ROUNSEWELL, M., SPANGENBERG, J., OMANN, I., STOCKER, A., CARTER, T.R., FRONZEK, S., JYLHÄ, K. & F.-W. BADECK (2005): Development, provision and application of the ALARM scenarios. – unpubl., distributed internally within the ALARM project.

ROOT, T.L., PRICE, J.T., HALL, K.R., SCHNEIDER, S.H., ROSENZWEIG, C. & J.A. POUNDS (2003): Fingerprints of global warming on wild animals and plants. *Nature* **421**: 57-60.

RÜBEL, E. (1930): Pflanzengesellschaften der Erde. – Huber, Bern.

SAKAI, A. (1982): Freezing resistance of ornamental trees and shrubs. *J. Amer. Soc. Hort. Sci.* **107**: 572-581.

SAKAI, A. & W. LARCHER (1987): Frost survival of plants. Ecological Studies 62. – Springer, Berlin.

SCHIMPER, A.F.W. (1935): Pflanzengeographie auf physiologischer Grundlage. 3rd ed. – Fischer, Jena.

SCHMID, E. (1939): Die Stellung Insubriens im Alpenbereich. *Verh. Schweiz. Naturforsch. Ges.:* 64-65.

SCHMID, E. (1956): Flora des Südens. – Rascher, Zürich.

SCHMITHÜSEN, J. (1976): Atlas zur Biogeographie. – Mannheim, Bibliographisches Institut.

SCHMUCKER, T. (1942): Die Baumarten der nördlich-gemäßigten Zone und ihre Verbreitung. *Silvae Orbis* **4**: 1-250.

SCHRÖTER, C. (1936): Flora des Südens. – Rascher, Zürich.

SCOLES, R. & N. VAN BREEMEN (1997): The effects of global change on tropical ecosystems. *Geoderma* **79**: 9-24.

SÖHLKE, G. (2006): Aktuelle und potenzielle Verbreitung der Lorbeer-Kirsche *Prunus laurocerasus* L. in Deutschland und angrenzenden Gebieten. – Diplomarbeit, Institut für Geobotanik, Universität Hannover.

STAUB, M. (1905): Die Geschichte des Genus *Cinnamomum. Math. Nat.wiss. Ber. Ungarn* **19**: 13-30.

SVENNING, J.-C. & F. SKOV (2004): Limited filling of the potential range in European tree species. *Ecol. Lett.* **7**: 565-573.

SYKES, M.T., PRENTICE, I.C. & W. CRAMER (1996): A bioclimatic model for the potential distributions of north European tree species under present and future climates. – *J. Biogeogr.* **23**: 203-223.

THOMAS, C.D., BODSWORTH, E.J., WILSON, R.J., SIMMONS, A.D., DAVIES, Z.G., MUSCHE, M. & L. CONRADT (2001): Ecological and evolutionary processes at expanding range margins. *Nature* **411**: 577-581.

THUILLER, W., LAVOREL, S., ARAÚJO, M.B., SYKES, M.T. & I.C. PRENTICE (2005): Climate change threats to plant diversity in Europe. *Proc. Nat. Acad. Sci. USA* **102**: 8245-8250.

TOMBAL, P. (1972): Etude phytocoenologique et esquisse macrobiocoenotique du proclimax forestier (Ilici-Fagetum) des Beaux-Monts de Compiégne (Oise-France). *Bull. Soc. Bot. Nord de la France* **25**: 19-29.

WALTER, H. & H. STRAKA (1970): Arealkunde. Floristisch-historische Geobotanik –Ulmer, Stuttgart.

WALTER, H. & S.-W. BRECKLE (1999): Vegetation und Klimazonen. – Ulmer, Stuttgart.

WALTHER, G.-R. (1999): Distribution and limits of evergreen broad-leaved (laurophyllous) species in Switzerland. *Bot. Helv.* **109**: 153-167.

WALTHER, G.-R. (2000): Climatic forcing on the dispersal of exotic species. –*Phytocoenologia* **30**: 409-430.

WALTHER, G.-R. (2001): Laurophyllisation - a sign of a changing climate? In: BURGA, C.A. & A. KRATOCHWIL (eds.): Biomonitoring: General and applied aspects on regional and global scales. Tasks for vegetation science 35. pp. 207-223. – Kluwer Academic Publishers, Dordrecht.

WALTHER, G.-R. (2002): Weakening of climatic constraints with global warming and its consequences for evergreen broad-leaved species. *Folia Geobot.* **37**: 129-139.

WALTHER, G.-R. (2003): Are there indigenous palms in Switzerland? *Bot. Helv.* **113**: 159-180.

WALTHER, G.-R., BURGA, C.A. & P.J. EDWARDS (eds.) (2001): 'Fingerprints' of climate change – Adapted behaviour and shifting species ranges. – Kluwer Academic/Plenum Publishers, New York.

WALTHER, G.-R., BERGER, S. & M.T. SYKES (2005): An ecological 'footprint' of climate change. *Proc. R. Soc. Lond. B* **272**: 1427-1432.

WALTHER, G.-R., GRITTI, E., BERGER, S., HICKLER, T., TANG, Z. & M.T. SYKES (2007): Palms tracking climate change. *Global Ecol. and Biogeogr.* Online publication DOI:10.1111/j.1466-8238.2007.00328.x

WARD, M.D. & R.F. LABISKY (2004): Post-dispersal germination success of native black gum (*Nyssa sylvatica*) and introduced camphor tree (*Cinnamomum camphora*) in Florida, USA. *Nat. Areas J.* **24**: 341-344.

WOHLGEMUTH, T., BOSCHI, K. & P. LONGATTI (2006): Swiss Webflora. – WSL. URL: http://www.wsl.ch/land/products/webflora/welcome-de.ehtml.

WOODWARD, F.I. (1987): Climate and plant distribution. – Cambridge University Press, Cambridge.

WOODWARD, F.I., LOMAS, M.R. & C.K. KELLY (2004): Global climate and the distribution of plant biomes. *Philos. Trans. R. Soc. Lond. B* **359**: 1465-1476.

ZELLER, O. (1951): Über die Assimilation und Atmung der Pflanze im Winter bei tiefen Temperaturen. *Planta* **39**: 500-526.

6 General discussion

The mean surface temperature in Europe increased by 0.95°C in the course of the 20th century. The rise in mean annual temperature was mainly due to higher winter temperatures and with an increasing warming rate towards the end of the century (IPCC 2001, EEA 2004). After the year 2000 even warmer years followed, and 11 of the 12 last years (1995-2006) rank among the 12 warmest years in the instrumental record on the global scale (IPCC 2007).

Ecological responses to recent climate change have been documented from different taxonomic groups and from different ecosystems around the world (e.g. WALTHER et al. 2002, PARMESAN & YOHE 2003, ROOT et al. 2003). Animals shift their ranges poleward and respond with phenological changes, as well as changes in their behaviour (e.g. FORCHHAMMER et al. 1998, POST et al. 1999, THOMAS & LENNON 1999, HILL et al. 2002, CRICK 2004, PARMESAN 2006). Plants have also been documented to be affected by climate change (e.g. WALTHER 2004). For instance, altitudinal upward shifts of mountain plants were detected in the European Alps (e.g. GRABHERR et al. 1994, WALTHER et al. 2005, PAULI et al. 2006), also changes in plant phenology in Europe have been documented (e.g. FITTER & FITTER 2002, MENZEL et al. 2006).

The process of evergreen broad-leaved species spreading into deciduous forests in the southernmost part of Switzerland has been documented since the 1980s (GIANONI et al. 1988, KLÖTZLI et al. 1996, CARRARO et al. 1999, 2001, WALTHER 2000, 2002). Whereas previous work mainly focused on the area surrounding the lakes in southern Switzerland (northern part of Lago Maggiore and northwestern part of Lago di Lugano, see also fig. 3.1a), this knowledge was updated based on various local reports and new field data, and the study area was extended to the whole Insubrian region (s. l.), including also the area along the Italian lakes (see chapter 3 "Distribution of evergreen broad-leaved woody species in Insubria in relation to bedrock and precipitation"). Numerous evergreen broad-leaved species are still present in deciduous forests close to settlements along the lakes in southern Switzerland and also in northern Italy. Throughout the whole Insubrian region, evergreen broad-leaved species are still restricted to southern slopes on lower latitudes, though the expansion process has continued and become more independent of the garden- and park-populations (see also chapter 4 "Palms tracking climate change"). WALTHER (2000, 2002) pointed out ameliorated climatic conditions due to recent climate change as a required prerequisite for the observed spread and naturalisation of evergreen broad-leaved species. The current investigation of the distribution of evergreen broad-leaved species in the whole Insubrian region after subsequent years with mild winters supports this conclusion, although further ecological parameters are additionally playing a role in determining the species distribution and abundance on the local scale (see chapter 3).

Many of the naturalising species, especially the most cold-hardy ones, are frequently cultivated in other parts of Europe. Rising winter temperatures on the continental scale consequently lead to the question whether this phenomenon is restricted to the Insubrian region or if a similar process has been initiated on a larger geographical scale. Historical and current distribution data on the most cold-hardy evergreen broad-leaved woody species occurring in Europe was analysed, concerning introduced as well as native species (chapter 2 "An ecological 'footprint' of climate change", chapter 4 "Palms tracking climate change" and chapter 5 "Bioclimatic limits and range shifts of cold-hardy evergreen broad-leaved species at their northern distributional limit in Europe"). Range shifts of single evergreen broad-leaved species at their northern range margins in Europe were detected, revealing that the spread of evergreen broad-leaved species is not restricted to the Insubrian region, although the most pronounced expansion has taken place in this region.

6.1 Temperature increase – the responsible driver for range shifts?

6.1.1 *Ilex aquifolium*

Not only on the global scale is a warming trend evident. As outlined in chapter 2, "An ecological footprint of climate change", a trend towards warmer winter temperatures was pinpointed in southern Scandinavia and Germany, based on long-term observational temperature data from climate stations.

Data from these climate stations was closely linked to the distribution of *Ilex aquifolium* in the work of IVERSEN (1944). Using the same methodology, the relationship between temperature and distribution of *Ilex aquifolium* was confirmed 60 years later, although a northward shift of the species range boundary and a synchronous climatic shift towards milder winter temperatures has taken place. The observed north- and north-eastward range expansion tracks the temperature increase measured at the local climate stations.

The range of *Ilex aquifolium* has shifted northward in Norway, where it is distributed along the west coast, currently as far north as 63° N (see chapter 3). Additionally, new occurrences further inland have been reported from the northern range margin (SALVESEN 1996). The historical distribution limit of *Ilex aquifolium* ran through Denmark, excluding the easternmost parts of the country as well as the island of Bornholm. The current distribution boundary is following the southern coast of Sweden, with the north-easternmost occurrences on the islands of Öland and Gotland. New occurrences beyond the former eastern range margin have also been documented in Germany (BERGER 2003, REHSE 2007). In the past, the 0°C isothermal line of January ran parallel with the northern boundary of *Ilex*' distribution. The isothermal line has shifted correspondingly with the recent range shift and the relationship still remains consistent.

Furthermore, the range shift of *Ilex aquifolium* corresponds well with the range predicted using the bioclimatic model STASH (SYKES *et al.* 1996), also supporting the conclusion that climate is the responsible driver for the observed range shift.

Other studies already linked species range shifts to the general warming trend observed in the last decades (e.g. GRABHERR *et al.* 1994, PARMESAN *et al.* 1999, THOMAS & LENNON 1999, HILL *et al.* 2002, CROZIER 2003). However, here the range shift of *Ilex aquifolium* is not only related to a global warming trend, but to local climate data, at the same time covering a large area, providing a higher confidence that climate is the responsible driver.

In the study of IVERSEN (1944) two other evergreen broad-leaved species, *Hedera helix* and *Viscum album* were also considered applicable as climate indicators. The altitudinal range of *Viscum album* ssp. *austriacum* is shifting towards higher elevations in the European Alps as a consequence of milder winters. The current upper range limit of *Viscum album* ssp. *austriacum* is ca. 200 m higher up than it was 100 years ago (DOBBERTIN *et al.* 2005). *Hedera helix* has two separate and different growth forms. *Hedera helix*, when creeping on the ground, is protected from winter cold by snow and may thus survive colder winters than when it is climbing up trees (ANDERGASSEN & BAUER 2002). In central European deciduous forests, an observed change of the growth form of *Hedera helix* from creeping along the ground to climbing up trees has likewise been linked to recent climate change (DIERSCHKE 2005).

6.1.2 *Trachycarpus fortunei*

One of the introduced species that have colonised deciduous forests in southern Switzerland is the palm *Trachycarpus fortunei,* native to Southeast Asia. Chapter 4 "Palms tracking climate change" outlines how this palm is establishing subspontanous populations in southern Switzerland, but also in other parts of Europe with mild winter conditions. Escaped *Trachycarpus fortunei* palms have been recorded in France as far north as Brittany as well as on the islands of the English Channel, and rejuvenation within gardens have been observed in southern England and recently also in southwest Germany.

The climatic conditions of the native range were explored in order to identify the bioclimatic limits of *Trachycarpus fortunei*. Winter temperature, *i.e.* January mean temperature, and the yearly growing degree days (GDD$_5$) were identified as key temperature parameters limiting the range of this species towards higher latitudes and higher altitudes. The threshold January mean temperature of + 2.2°C realised in the native range was compared with long-term measurements from the climate station Lugano. Although the palm has been cultivated for centuries in the Lugano region, the threshold value of + 2.2°C January mean temperature was not exceeded for longer periods until the end of the 20th century. A reproductive, sustainable population of *Trachycarpus fortunei* was first established in this time span. The results of the STASH bioclimatic model, which allowed a geographic projection of the potential suitable habitat in Europe in different periods of the past and present, also show that periods with a continuous suitable climate have not occurred in Ticino until the last few decades. This strongly suggests that an ameliorated climate, especially warmer and shorter winter seasons, was the essential prerequisite for *Trachycarpus fortunei* to become locally established in Ticino. Furthermore, the improved understanding of the importance of sublethal frost incidents as outlined in chapter 4 supports this conclusion.

The numerous further observations of rejuvenating *Trachycarpus fortunei* in other parts of Europe can be interpreted as an early stage of a possible similar invasion process, which may in the long run lead to further sustaining populations as conditions become increasingly suitable, assuming that climate warming will continue.

Worldwide, *Trachycarpus fortunei* is not the only palm expanding northward, though it is naturalising also in the USA, in Australia and New Zealand. In south-western North America the indigenous species *Washingtonia filifera* is expanding northward beyond its former range (CORNETT 1991). So does also *Sabal mexicana* in Texas (LOCKETT 2004) and *Sabal minor* in parts of Tennessee and other areas of the South-East where it is not native (D. FRANCKO, pers. comm., see also BJORHOLM *et al.* 2005).

6.1.3 *Prunus laurocerasus*

A similar methodology was applied to derive limiting parameters of the evergreen broad-leaved shrub *Prunus laurocerasus* (SÖHLKE 2006). The cherry laurel, *Prunus laurocerasus*, is native to the Balkans and the coast of the Black Sea and Caspian Sea, and it is frequently cultivated in central Europe as an ornamental shrub. Winter temperature and length of the growing season were determined to be important climatic parameters limiting the distribution of *Prunus laurocerasus*. The limiting January mean temperature based on the native range was estimated to be -1.2°C.

Prunus laurocerasus additionally needs a sufficiently long growing season in order to produce viable seeds (ADAMOVIĆ 1909, SÖHLKE 2006). The potential distribution of *Prunus laurocerasus* based on these parameters overlaps well with the observed occurrences. For further details see chapter 5 "Bioclimatic limits and range shifts of cold-hardy evergreen broad-leaved species at their northern distributional limit in Europe".

6.1.4 Limiting temperature parameters

Winter temperature plays an important role in limiting the distribution of evergreen broad-leaved vegetation towards the poles (BOX 1981, SAKAI & LARCHER 1987, WOODWARD 1987, WALTER & BRECKLE 1999). Although the occurrence of minimum temperatures as low as -15°C is an important threshold value for evergreen broad-leaved vegetation in general, the most cold-hardy evergreens extend their ranges into areas where these values are reached from time to time, such as *Ilex aquifolium*. Temperature plays an important role limiting the northern range of both native and introduced evergreen broad-leaved species in Europe, though the threshold values and biological impacts of different temperature parameters are species specific as discussed in detail in chapter 3 "Bioclimatic limits and range shifts of cold-hardy evergreen broad-leaved species at their northern distributional limit in Europe" and in chapter 4 "Palms tracking climate change".

Cold-hardiness, also with regard to extreme temperatures, differs considerably between the species. Some experimentally determined values of resistance to extremely low temperatures are compiled in table 5.1. Furthermore, cold hardiness differs depending on stage of life cycle and preadaptation. Different tissues are not equally resistant to low temperatures, for instance may over wintering propagative organs be especially vulnerable to low temperatures, as SAKAI & LARCHER (1987) experimentally showed for *Laurus nobilis*. In the field, duration of extreme events and accompanying weather conditions (e.g. snow cover) also influence the impact of severe frost incidents (LARCHER 1994, BRUNOLD *et al.* 1996).

However, not only extremes are limiting plant growth and distribution; specific biological traits make species vulnerable to different climatic phenomena. In addition to low winter temperature, expressed as minimum or mean temperatures, which plays a role in limiting the northern range of all species regarded in this study, other temperature parameters may additionally limit the species' persistence at northern range margins, such as the length of the growing season and total summer warmth.

Whereas the survival and regeneration of *Ilex aquifolium* is mainly restricted by severe winters, *Prunus laurocerasus* additionally needs a sufficiently long growing season in order to produce viable seeds (ADAMOVIĆ 1909, SÖHLKE 2006). Vegetative growth and survival of cultivated individuals is possible beyond the areas where reproduction is possible. This has been observed on high altitudes in the native range, and in Europe, where *Prunus laurocerasus* is cultivated as far north as southern Scandinavia. *Prunus laurocerasus* sheds its fruits in autumn, in contrast to species like *Trachycarpus fortunei* and *Ilex aquifolium,* which keeps their fruits all winter. The fruits of *Ilex aquifolium* are ripening throughout the following growing season (CALLAUCH 1983); a similar cumulative ripening process is not possible for the seeds of *Prunus laurocerasus*. Thus, length of the growing season restricts reproduction, and may restrict *Prunus laurocerasus* from extending as far north as *Ilex aquifolium*, in spite of its ability to survive short term frost incidents with temperatures as low as -20 °C (BÄRTELS 1991) to -24 °C (DIRR & LINDSTROM 1990).

Species with subtropical distribution in particular, like *Trachycarpus fortunei*, may require sufficiently warm summers to assure continuous growth progress. Warm summers allow this palm to accumulate enough biomass to regenerate damaged leaves and thereby to compensate for sublethal frost damage, as discussed in detail in chapter 4 "Palms tracking climate change". Very warm summers may hence increase the tolerable frost damage in winter to some degree, and so increase cold-hardiness, though winter temperatures must still exceed a minimum threshold value.

6.1.5 Climate variables

For the most part, average temperature variables were applied in this study, although minimum temperatures of extreme cold events are important for the mechanistic understanding of climatic limits. Minimum temperatures are correlated with the mean temperature of the coldest month and may be used as a surrogate for absolute minimum temperatures (DAHL 1998, PRENTICE *et al.* 1992). Mean temperature of the coldest month is also more robust to local deviations due to microclimatic conditions than the absolute minimum temperature (DAHL 1998). Hence it is more representative of the regional macroclimate and consequently applicable for large-scale investigations. The temperature of the coldest month has been successfully applied in other studies analysing bioclimatic limits (e.g. IVERSEN 1944, HINTTIKKA 1963), as well as in bioclimatic modelling (e.g. PRENTICE *et al.* 2002, SYKES *et al.* 1996, WOODWARD *et al.* 2004).

The relationship between extreme cold events and January mean temperatures has also been tested at the eastern distribution boundary of *Ilex aquifolium* in Germany. A distinct decrease in extreme cold events in eastern Germany in the last decades was accompanied with an eastward shift of the 0°C isothermal line of January. In the respective regions new occurrences of *Ilex aquifolium* were recorded (REHSE 2007).

6.2 Detection and verification of vegetation shifts with bioclimatic models

Results from bioclimatic modelling was included in several parts of this study and proved to be a sophisticated tool to obtain geographical projections of certain combinations of climatic limits. Furthermore modelling served to extract and visualise impacts of climate, separated from other influences, and supplemented the empirical field observations as a theoretical approach.

Most studies applying bioclimatic modelling approaches to the European flora do not regard introduced species. The results of this study verified with ground-truth data that bioclimatic modelling may also be successfully applied to introduced species, if sufficient data from the native range can be provided, which, however, is not always given.

Bioclimatic modelling based on and validated with field data may be applied to detect areas where climate change is likely to induce certain changes, and thus where field surveys are particularly important. Verification of bioclimatic model output in the field may on the other hand reveal effects that are not implemented in models, and remains necessary to attribute confidence to modelling results. Several provisions, preconditions and limitations must be understood and taken into account (BOTKIN *et al.* 2007) interpreting the modelling results. Hence, field surveys remain the key to address the whole complexity of vegetation shifts induced by climate change, as many processes are still not completely understood, although bioclimatic modelling may be useful as a supplementary tool, addressing clearly defined questions.

6.3 Native vs. introduced evergreen broad-leaved species

As outlined in the previous sections, native as well as introduced evergreen broad-leaved species are expanding their ranges is central and northern Europe. An example of a native species' range expansion is the northward range shift of *Ilex aquifolium*. The new northernmost occurrences mostly originated from native populations. In some cases, however, especially those concerning the eastern distribution boundary, new individuals were considered to be escapees from planted garden individuals. IVERSEN (1944) also included a category named "*Ilex* strayed into woods from gardens" in his study. Hence, such sub-spontaneous occurrences were also included in the findings of the original study and in agreement with the methodology of IVERSEN (1944). In Norway, a strict distinction of separate "wild" and "cultivated" populations is difficult, as there has been a long-term exchange between these populations. On the one hand, most cultivated individuals originate from the indigenous populations. On the other hand, the planted individuals acted as seed sources contributing to the recovery of the forest populations after severe winters (SALVESEN 1996, see also BERGER 2003).

Whereas *Laurus nobilis* is considered native to northern Italy, the species is regarded as an archeophyt in southern Switzerland, where it has spread from plantings in many locations. These spreadings, too, are a geographical extension of the native range. Hence, the range shifts of *Ilex aquifolium* and *Laurus nobilis* are effectively extensions of their historical ranges. These species are additionally favoured by direct human impacts, as they are cultivated beyond their range boundaries. These planted individuals may serve as additional seed sources, and hence accelerate the expansion. Cultivation provides opportunities for the species to keep pace with the rate of climate change by reducing the time-lag that may be due to, for example, dispersal limitations (SVENNING & SKOV 2004, cf. also POTT 1990, LEEMANS 1996) and/or interrupted migration routes (SKOV & SVENNING 2004) thus allowing a species to occupy its new potential range immediately.

Laurus nobilis and *Quercus ilex* do also occur in Southern England and Ireland, where they are clearly originating from individuals naturalised from plantings.

Regarding the origin of the species considered in this study, a transition from native to exotic seems useful, rather than a strict separation of native and exotic species: From *Ilex aquifolium*, considered native in northern Europe, via *Prunus laurocerasus*, native to Europe in the Tertiary, though isolated in its current distribution area due to glaciations throughout the Quaternary, to species from other parts of the world that first reached Europe due to direct human action, e.g. *Trachycarpus fortunei, Cinnamomum glanduliferum* .

Regardless of the origin of the investigated species, their ecological requirements overlap to a considerable extent, due to their common functional type as broad-leaved evergreens, as reflected in fig. 5.2.

6.4 Impact of precipitation and bedrock

6.4.1 Impact of precipitation

In previous sections the specificity of temperature requirements has been highlighted. Further ecological demands are species specific as well, for instance do precipitation requirements differ considerably between the species investigated in this study, and may even limit their distribution. In Europe this is predominantly the case at their southern range margins. Concerning the distribution and bioclimatic limits of the species, the influence of precipitation must therefore also be taken into account.

Within the Insubrian region (*s. l.*) from Lago d'Orta in the West to Lago di Garda in the East a pronounced precipitation gradient is realised. Precipitation is ranging from ca. 2000 mm per year in the region of Lago d'Orta to less than 1000 mm per year at the southern end of Lago di Garda. The effect of this gradient in the yearly amount of rainfall is further amplified by the seasonal pattern with summer rain in the west and 1 to 3 months of summer drought in the east (REISIGL 1996, SCHWARB *et al.* 2002). The investigation of species composition of evergreen broad-leaved species and their degree of establishment in different study sites along the precipitation gradient revealed distinct differences in the composition, distribution and abundance of evergreen broad-leaved species between the study sites (see chapter 3).

Species like *Cinnamomum glanduliferum* and *Trachycarpus fortunei* originate from areas with subtropical, humid climate and are relatively sensitive to drought. In the Insubrian region they are restricted to the western part of the area with high precipitation throughout the year. On the other hand, species like *Quercus ilex* and *Viburnum tinus* are prevailing in the eastern part of the area, where summer drought is more pronounced and annual precipitation is considerably lower. Some species are consistent throughout the whole area, such as *Laurus nobilis* and *Ilex aquifolium*. The more drought tolerant species are also occurring in (sub-) Mediterranean regions, where summer drought limits the distribution of e.g. *Cinnamomum glanduliferum, Trachycarpus fortunei* and *Prunus laurocerasus*.

Hence, species can be assigned to a precipitation gradient according to their requirements as outlined in detail in chapter 3. This gradient is furthermore reflected in species' leaf morphology; the laurophyllous leaf type occurs under Insubrian / atlantic climatic conditions, whereas a higher degree of sclerophylly occurs towards more Mediterranean areas.

In the literature, two bioclimatic and floristic Insubrian subtypes are distinguished for the native flora (e.g. OBERDORFER 1964, REISIGL 1996). These subtypes are also reflected in the distribution of introduced evergreen broad-leaved species and in the species composition of the study sites.

6.4.2 Impact of bedrock

Differences in the geological bedrock further support the distinction between the western and the eastern part of the Insubrian region. Whereas siliceous substrates are dominant in the western part of the area, the region of Lago di Garda and Lago d'Iseo in the east is dominated by calcareous bedrock. Edaphic and geological parameters are important ecological factors limiting the distribution of individual species on regional to local scales (e.g. SITTE *et al.* 2002, POTT 2005).

Some of the investigated species show clear preferences for one of the substrates, as discussed in chapter 3. Whereas *Cinnamomum glanduliferum* is restricted to silicious soils, *Quercus ilex* is only occurring on calcareous soils in the investigated area. In the western part, where precipitation is high, siliceous bedrock is predominant. The decay of organic matter is relatively slow and allows the accumulation of a relatively thick humus layer (BLASER 1973). In contrast, the calcareous underground in the eastern part of the area provides shallow soils with a lower water storage capacity, increasing the risk of drought stress. With increasing risk of drought stress the laurophyllous species become less abundant, and they are increasingly replaced by sclerophyllous, (sub-)Mediterranean species, as outlined in the previous section. In general, this study shows that the effect of the precipitation gradient in the study area is modified, mostly amplified, by the geological bedrock.

6.5 Evergreen broad-leaved species – indicators of climate change?

Superposing potential ranges of several evergreen broad-leaved species revealed distinct similarities in the centre of the potential distribution of the species, although species limits are specific as previously outlined. The potential species distributions overlap in substantial parts of central Europe, with a gradually decreasing number of species' ranges towards the northeast (see fig 5.2 and chapter 5).

Both the southernmost part of Switzerland and southern Great Britain have especially favourable winter temperatures, compared with other parts of central Europe where deciduous forests occur. In these regions the potential distributions of several species overlap. This is in agreement with observational data, which confirm that the species do actually occur in these regions. Even more evergreen broad-leaved species, that are otherwise rare in central Europe, concentrate in these regions. Updated distribution data from France are scarcer, but data from MUSÉUM NATIONAL D'HISTOIRE NATURELLE (2003-2006) indicates that many of the same species are becoming more frequent in atlantic regions of the country. The most cold-hardy evergreen species are also occurring and spreading in Germany, such as *Ilex aquifolium*, *Prunus laurocerasus* and *Mahonia aquifolium* (BERGER 2003, SÖHLKE 2006, BFN 2007, REHSE 2007). Furthermore, single occurrences of other species have also been recorded, e.g. *Aucuba japonica, Viburnum rhytidophyllum* (BRANDES 2003, SCHMITZ et al. 2004).

The most obvious effect of rising winter temperatures is a reduced risk of lethal frost incidents, which allows survival beyond the former range boundary. Additionally, the frequency of sub-lethal frosts decreases and sub-lethal damage, which must be compensated by replacement of lost biomass, will decrease. Hence, survival in formerly unsuitable areas may become possible under climatic warming. Evergreen broad-leaved species are particularly favoured by milder winter temperatures, as they are able to profit from positive net photosynthesis in periods with favourable climatic conditions even in winter (ZELLER 1951, FISCHER & FELLER 1994, OLIVEIRA & PEÑUELAS 2004), in contrast to deciduous trees which are leafless in the winter season.

Conclusions

The results of this study show that the cold-hardiest evergreen broad-leaved species are expanding on the European scale. Single species' range shifts have been linked to climate change, *i.e.* rising winter temperatures. Consequently, and with regard to the overlap of potential distribution areas of the species, it may be concluded that the same underlying process, *i.e.* climate change, is the responsible driver of the synchronous expansion of several evergreen broad-leaved species.

Within the group of evergreen broad-leaved species regarded in this study, different biological traits influence the distribution of the single species. However, they are all sensitive to low winter temperatures, though at different threshold values, and hence their ranges are limited by their specific cold-hardiness. This is further reflected in their expansion in response to the milder winters of the last decades and thus, they are suitable as indicators of recent climate change in central Europe.

6.6 Impacts on the vegetation level (Outlook)

Observed changes

The substantial overlap of the projected ranges of several evergreen broad-leaved species in southern Great Britain and southern Switzerland (as shown in Fig. 5.2) suggests a trend towards more evergreen broad-leaved species in deciduous forests of some European regions, as discussed in detail in chapter 5. Deciduous forest with evergreen understorey has been recognised as a separate vegetation type in southern Switzerland (DELARZE et al. 1999). In Great Britain, small evergreen forests and scrub stands (often with *Quercus ilex*) have been described as recently generated vegetation types with no counterpart in the native vegetation, whereas *Rhododendron* thickets and understoreys with a high share of further evergreen species, e.g. *Prunus laurocerasus*, are matched in the native vegetation by evergreen thickets of *Ilex aquifolium* (PETERKEN 2001). Hence, not only single species are shifting their ranges; in fact, an effect on the vegetation level is also evident, with several evergreen broad-leaved species forming a new community in former deciduous forests. This may not only lead to distinct changes in the diversity and physiognomy of existing forest communities, but also to changes in forest ecosystem processes, e.g. carbon cycling and water balance. So far, these ecosystem level consequences have not been addressed, but they could be subject to further studies on the ecological impacts of the spread and establishment of evergreen broad-leaved species.

As discussed in chapter 5, the emerging vegetation type symbolises a revival of certain elements of the Tertiary vegetation of Europe. But as most species common in the Tertiary are extinct in central Europe, distinct differences are apparent. The observed species belong to genera well-known from the Tertiary, but are introduced from other parts of the world due to human action. Furthermore, the combination of different introduced evergreen broad-leaved species with species characteristic of present-day European deciduous forests does not resemble the European Tertiary vegetation.

Possible future changes

Vegetation modelling has projected range shifts both by single species and by biomes due to climate change (e.g. PRENTICE *et al.* 1992, BAKKENES *et al.* 2002, WOODWARD *et al.* 2004, THUILLER *et al.* 2005, OHLEMÜLLER *et al.* 2006). However, the results presented here show that responses to climate change are species-specific, suggesting that climate change in the longer term will lead to a reorganisation of existing communities rather than to synchronous shifts of existing vegetation units as a whole. Furthermore, models predict that combinations of climate variables will arise that do not currently exist anywhere on the globe (SAXON *et al.* 2005, WILLIAMS *et al.* 2007). These new climates are expected to cause ecosystem reshuffling as individual species, constrained by different environmental factors, respond differently (FOX 2007), resulting in so-called "no-analog" ecosystems, which are also known from the geological record in combination with no-analog climates of the past (OVERPECK *et al.* 1992, JACKSON & WILLIAMS 2004). However, vegetation and ecological communities in general are also characterised by their inherent interactions, and hence accounting only for climate effects may underestimate the magnitude of ecological change (cf. FOX 2007).

Recent climate change has favoured the evergreen broad-leaved plant functional type within deciduous forests. With continued warming this process is likely to proceed and to induce changes in the composition and structure of temperate deciduous forests in various parts of central Europe. In addition to climatic changes, other human activities also cause changes to the native vegetation, for example through the introduction of exotic species. The combination of several different anthropogenic influences and climate change has facilitated this vegetation type which may – given continued climate change – symbolise a first step towards a no-analog plant community in Europe.

6.7 References

ADAMOVIĆ, L. (1909): Die Vegetationsverhältnisse der Balkanländer. – Engelmann, Leipzig.

ANDERGASSEN, S. & H. BAUER (2002): Frost hardiness in the juvenile and adult life phase of Ivy (*Hedera helix* L.). *Plant Ecol.* **161**: 207-213.

BAKKENES, M., ALKEMADE, J.R.M., IHLE, F., LEEMANS, R. & J.B. LATOUR (2002): Assessing effects of forecasted climate change on the diversity and distribution of European higher plants for 2050. *Glob. Chang. Biol.* **8**: 390-407.

BÄRTELS, A. (1991): *Prunus laurocerasus*, die Lorbeerkirsche. *Baumschulpraxis* **8**: 324-327.

BERGER, S. (2003): *Ilex aquifolium* – Bioindikator für Klimaveränderung? – Diplomarbeit, Institut für Geobotanik, Universität Hannover.

BfN (2007): Floraweb. – BfN, Bonn. URL: http://www.floraweb.de, 30.05.07.

BJORHOLM, S., SVENNING, J.-C., SKOV, F. & H. BALSLEV (2005): Environmental and spatial controls of palm (Arecaceae) species richness across the Americas. *Global Ecol. Biogeogr.* **14**: 423-429.

BLASER, P. (1973). Die Bodenbildung auf Silikatgestein im südlichen Tessin. *Mitt. Schweiz. Anst. forstl. Vers.wes.* **49**: 254-340.

BOTKIN, D.B., SAXE, H., ARAUJO, M.B., BETTS, R., BRADSHAW, R.H.W., CEDHAGEN, T., CHESSON, P., DAWSON, T.P., ETTERSON, J.R., FAITH, D.P., FERRIER, S., GUISAN, A., HANSEN, A.J., HILBERT, D.W., LOEHLE, C., MARGUELS, C., NEW, M., SOBEL, M.J. & D.R.B. STOCKWELL (2007): Forecasting the effects of global warming on biodiversity. *Bioscience* **57** (3): 227-236.

BOX, E. O. (1981): Macroclimate and plant forms: An introduction to predictive modelling in phytogeography. Tasks for vegetation science 1. – Junk Publishers, The Hague.

BRANDES, D. (2003): Die aktuelle Situation der Neophyten in Braunschweig. *Braunschweiger Naturkundliche Schriften* **6** (**4**): 705-760.

BRUNOLD, C., RÜEGSEGGER, A. & R. BRÄNDLE. (1996): Stress bei Pflanzen. – Haupt, Bern.

CALLAUCH, R. (1983): Untersuchungen zur Biologie und Vergesellschaftung der Stechpalme *(Ilex aquifolium)*. – Dissertation, Universität des Landes Hessen, Kassel.

CARRARO, G., GIANONI, G., MOSSI, R., KLÖTZLI, F. & G.-R. WALTHER (2001): Observed changes in vegetation in relation to climate warming. In: BURGA, C.A. & KRATOCHWIL, A. (eds.): Biomonitoring: General and applied aspects on regional and global scales. Tasks for vegetation science 35. pp. 195-205. – Kluwer academic publishers, Dordrecht.

CARRARO, G., KLÖTZLI, F., WALTHER, G.-R., GIANONI, P. & R. MOSSI (1999): Observed changes in vegetation in relation to climate warming. Final Report NRP 31. – vdf Hochschulverlag, Zürich.

CORNETT, J.W. (1991): Population dynamics of the palm, *Washingtonia filifera*, and global warming. *San Bernardino County Museum Association Quarterly* **39**: 46-47.

CRICK, H. (2004): The impact of climate change on birds. *Ibis* **146** (Suppl.1): 48-56.

CROZIER, L. (2003): Winter warming facilitates range expansion: cold tolerance of the butterfly *Atalopedes campestris*. *Oecologia* **135**: 648-656.

DAHL, E. (1998): The Phytogeography of Northern Europe. – Cambridge University Press, Cambridge.

DELARZE, R., GONSETH, Y. & P. GALLAND (1999): Lebensräume der Schweiz. – Ott Verlag, Thun.

DIERSCHKE, H. (2005): Laurophyllisation - auch eine Erscheinung im nördlichen Mitteleuropa? Zur aktuellen Ausbreitung von *Hedera helix* in sommergrünen Laubwäldern. *Ber. Reinh.-Tüxen-Ges.* **17**: 151-168.

DIRR, M.A. & O.M. LINDSTROM (1990): Leaf and stem cold hardiness of 17 broadleaf evergreen taxa. *J. Environ. Horticult.* **8**: 71-73.

DOBBERTIN, M., HILKER, N., REBETEZ, M., ZIMMERMANN, N.E., WOHLGEMUTH, T. & A. RIGLING (2005): The upward shift in altitude of pine mistletoe (*Viscum album* ssp. *austriacum*) in Switzerland – the result of climate warming? *Int. J. Biometeorol.* **50**: 40-47.

EEA (2004): EUA Signale 2004. Aktuelle Informationen der Europäischen Umweltagentur zu ausgewählten Themen. – Amt für amtliche Veröffentlichungen der Europäischen Gemeinschaften, Luxemburg.

FISCHER, A. & U. FELLER (1994): Seasonal changes in the pattern of assimilatory enzymes and proteolytic activities in leaves of juvenile ivy. *Ann. Bot.* **74**: 389-396.

FITTER A.H. & R.S.R. FITTER (2002): Rapid changes in the flowering time in British plants. *Science* **296**: 1689-1691.

FORCHHAMMER, M., POST, E. & N.C. STENSETH (1998): Breeding phenology and climate. *Nature* **391**: 29-30.

FOX, D. (2007): Back to the No-Analog Future. *Science* **316**: 823-825.

GIANONI, G., CARRARO, G. & F. KLÖTZLI (1988): Thermophile, an laurophyllen Pflanzenarten reiche Waldgesellschaften im hyperinsubrischen Seenbereich des Tessins. *Ber. Geobot. Inst. ETH, Stiftung Rübel, Zürich* **54**: 164-180.

GRABHERR, G., GOTTFRIED, M. & H. PAULI (1994): Climate effects on mountain plants. *Nature* **369**: 448.

HILL, J.K., THOMAS, C.D., FOX, R., TELFER, M.G., WILLIS, S.G., ASHER, J. & B. HUNTLEY (2002): Responses of butterflies to twentieth century climate warming: implications for future ranges. *Proc. R. Soc. Lond.* B **269**: 2163-2171.

HINTTIKKA, V. (1963): Über das Grossklima einiger Pflanzenareale in zwei Klimakoordinaten dargestellt. *Ann. Bot. Soc. Zool.-bot. Fenn. Vanamo* **34 (5)**: 1-64.

IPCC (2001): Climate Change 2001: The scientific basis. Contribution of working group I to the third assessment report of the intergovernmental panel on climate change. – Cambridge University Press, Cambridge.

IPCC (2007): Climate Change 2007: The Physical Science Basis. – IPCC, Genf. URL: http://www.ipcc.ch, 15.06.07.

IVERSEN, J. (1944): *Viscum, Hedera* and *Ilex* as climatic indicators. *Geol. Fören. Förhandl.* **66**: 463-483.

JACKSON S.T. & J.W. WILLIAMS (2004): Modern analogs in Quartenary paleoecology: Here today, gone yesterday, gone tomorrow? *Annu. Rev. Earth Planet. Sci.* **32**: 495-537.

KLÖTZLI, F., WALTHER, G.-R., CARRARO, G. & A. GRUNDMANN (1996): Anlaufender Biomwandel in Insubrien. *Verh. Ges. Ökol.* **26**: 537-550.

LEEMANS, R. (1996): Biodiversity and global change. In: K.J. GASTON (ed.): Biodiversity – A Biology of Numbers and Difference. pp. 367-387. – Blackwell, Oxford.

LARCHER, W. (1994): Ökophysiologie der Pflanzen. –Ulmer, Stuttgart.

LOCKETT, L. (2004) The Sabal Palm: Restoring a species we didn't know we had (Texas). *Ecological Restoration* **22**: 137-138.

MENZEL, A., SPARKS, T.H., ESTRELLA, N., KOCH, E., AASA, A., AHAS, R., ALM-KÜBLER, K., BISSOLLI, P., BRASLAVSKA, O., BRIEDE, A., CHMIELEWSKI, F.M., CREPINSEK, Z., CURNEL, Y., DAHL, Å., DEFILA, C., DONNELLY, A., FILELLA, Y., JATCZAK, K., MÂGE, F., MESTRE, A., NORDLI, Ø., PEÑUELAS, J., PIRINEN, P., REMIŠOVÁ, V., SCHEIFINGER, H., STRIZ, M., SUSNIK, A., VAN VLIET, A.J.H., WIEGOLASKI, F.-E., ZACH, S. & A. ZUST (2006): European phenological response to climate change matches the warming pattern. *Glob. Chang. Biol.* **12**: 1969-1976.

MUSÉUM NATIONAL D'HISTOIRE NATURELLE (2003-2006): Inventaire national du Patrimonie naturel. URL: http://inpn.mnhn.fr, Version 02.04.07.

OBERDORFER, E. (1964): Der insubrische Vegetationskomplex, seine Struktur und Abgrenzung gegen die submediterrane Vegetation in Oberitalien und in der Südschweiz. *Beitr. naturk. Forsch. SW-Deutschl.* **23**: 141-187.

OHLEMÜLLER, R., GRITTI, E.S., SYKES, M.T. & C.D. THOMAS (2006): Quantifying components of risk for European woody species under climate change. *Glob. Chang. Biol.* **12**: 1788-1799.

OLIVEIRA, G. & J. PEÑUELAS (2004): Effects of winter cold stress on photosynthesis and photochemical efficiency of PSII of the Mediterranean *Cistus albidus* L. and *Quercus ilex* L. *Plant Ecol.* **175**: 179-191.

OVERPECK, J.T., WEBB, R.S. & T. WEBB (1992): Mapping eastern north-American vegetation change of the past 18 kA – No-Analogs and the future. *Geology* **20 (12)**: 1071-1074.

PARMESAN, C., RYRHOLM, N., STEFANESCU C., HILL, J.K., THOMAS, C.D., DESCIMON, H., HUNTLEY, B., KAILA, L., KULLBERG, J., TAMMARU, T., TENNET, W.J., THOMAS J.A. & M. WARREN (1999): Poleward shifts in geographical ranges of butterfly species associated with regional warming. *Nature* **399**: 579-583.

PARMESAN, C: (2006): Ecological and evolutionary responses to recent climate change. *Annu. Rev. Ecol. Syst.* **37**: 637-669.

PARMESAN, C. & G. YOHE (2003): A globally coherent fingerprint of climate change impacts across natural systems. *Nature* **421**: 37-42.

PAULI, H., GOTTFRIED, M., REITER, K., KLETTNER, C. & G. GRABHERR (2006): Signals of range expansions and contractions of vascular plants in the high Alps: observations (1994–2004) at the GLORIA master site Schrankogel, Tyrol, Austria. *Glob. Chang. Biol.* **12**: 1-10.

PETERKEN, G.F. (2001): Ecological effects of introduced tree species in Britain. *For. Ecol. Manage.* **141**: 31-42.

POST, E., PETERSON, R.O., STENSETH, N.C. & B.E. MCLAREN (1999): Ecosystem consequences of wolf behavioural response to climate. *Nature* **401**: 905-907.

POTT, R. (1990): Die nacheiszeitliche Ausbreitung und heutige pflanzensoziologische Stellung von *Ilex aquifolium* L. *Tuexenia* **10**: 497-512.

POTT, R. (2005): Allgemeine Geobotanik. – Springer, Berlin.

PRENTICE, I.C., CRAMER, W., HARRISON, S.P., LEEMANS, R., MONSERUD, R.A. & A.M. SOLOMON (1992): A global biome model based on plant physiology and dominance, soil properties and climate. *J. Biogeogr.* **19**: 117-134.

REHSE, A.-K. (2007): *Ilex aquifolium* – Verbreitung und deren Grenzen im östlichen Deutschland. – Diplomarbeit, Institut für Geobotanik, Universität Hannover.

REISIGL, H. (1996): Insubrien und das Gardaseegebiet – Vegetation, Florengeschichte, Endemismus. *Ann. Mus. civ. Rovereto* **11** (Suppl. II): 9-25.

ROOT, T.L., PRICE, J.T., HALL, K.R., SCHNEIDER, S.H., ROSENZWEIG, C. & J.A. POUNDS (2003): Fingerprints of global warming on wild animals and plants. *Nature* **421**: 57-60.

SAKAI, A. & W. LARCHER (1987): Frost survival of plants. Ecological studies 62. – Springer, Berlin.

SALVESEN, P.H. (1996): *Ilex aquifolium* L. in Norway. *Holly Society Journal* **14**: 10-18.

SAXON, E., BAKER, B., HARGROVE, W., HOFFMANN, F. & C. ZGANJAR (2005): Mapping environments at risk under different global climate change scenarios. *Ecol. Letters* **8**: 53-60.

SCHMITZ, G., G. KASPAREK & K. ADOLPHI (2004): *Aucuba japonica* Thunb. ex Murr. (Cornaceae) auf dem Weg der Naturalisation? *Floristische Rundbriefe* **37**: 5-9.

SCHWARB, M., DALY, C., FREI, C. & C. SCHÄR (2000): Mittlere jährliche Niederschlagshöhen im europäischen Alpenraum 1971-1990. – Hydrologischer Atlas der Schweiz, Bern.

SITTE, P., WEILER, E.W., KADEREIT, J.W., BRESINSKY, A. & C. KÖRNER (2002): Strasburger Lehrbuch der Botanik. – Spektrum, Heidelberg.

SKOV, F. & J.C. SVENNING (2004): Potential impact of climatic change on the distribution of forest herbs in Europe. *Ecography* **27**: 366-380.

SÖHLKE, G. (2006): Aktuelle und potenzielle Verbreitung der Lorbeer-Kirsche *Prunus laurocerasus* L. in Deutschland und angrenzenden Gebieten. – Diplomarbeit, Institut für Geobotanik, Universität Hannover.

SVENNING, J.-C. & F. SKOV (2004): Limited filling of the potential range in European tree species. *Ecol. Letters* **7**: 565-573.

SYKES, M.T., PRENTICE, I.C. & W. CRAMER (1996): A bioclimatic model for the potential distributions of north European tree species under present and future climates. *J. Biogeogr.* **23**: 203-223.

THOMAS, C.D. & J.J. LENNON (1999): Birds extend their ranges northwards. *Nature* **399**: 213.

THUILLER, W., LAVOREL, S., ARAÚJO, M.B., SYKES, M.T. & I.C. PRENTICE (2005): Climate change threats to plant diversity in Europe. *Proc. Nat. Acad. Sci. USA* **102**: 8245-8250.

WALTER, H. & S.-W. BRECKLE (1999): Vegetation und Klimazonen. – Ulmer, Stuttgart.

WALTHER, G.-R. (2000): Climatic forcing on the dispersal of exotic species. *Phytocoenologia* **30**: 409-430.

WALTHER, G.-R. (2002): Weakening of climatic constraints with global warming and its consequences for evergreen broad-leaved species. *Folia Geobot.* **37**: 129-139.

WALTHER, G.-R. (2004): Plants in a warmer world. *Perspect. Plant Ecol. Evol. Syst.* **6**: 169-185.

WALTHER, G.-R., POST, E., CONVEY, P., MENZEL, A., PARMESAN, C., BEEBEE, T.J.C., FROMENTIN, J.-M., HOEGH-GULDBERG, O. & F. BAIRLEIN (2002): Ecological responses to recent climate change. *Nature* **416**: 389-395.

WALTHER, G.-R., BEISSNER, S. & C.A. BURGA (2005): Trends in upward shift of alpine plants. *J. Veg. Sci.* **16**: 541-548.

WILLIAMS, J.W., JACKSON, S.T. & J.E. KUTZBACH (2007): Projected distributions of novel and disappearing climates by 2100 AD. *Proc. Nat. Acad. Sci. USA* **104**: 5738-5742.

WOODWARD, F.I. (1987): Climate and plant distribution. – Cambridge University Press, Cambridge.

WOODWARD, F.I., LOMAS, M.R. & C.K. KELLY (2004): Global climate and the distribution of plant biomes. *Philos. Trans. R. Soc. B* **359**: 1465-1476.

ZELLER, O. (1951): Über die Assimilation und Atmung der Pflanze im Winter bei tiefen Temperaturen. *Planta* **39**: 500-526.

Acknowledgements

This study was carried out at the Institute of Geobotany of the Gottfried Wilhelm Leibniz University of Hanover, where, under leadership of Prof. Dr. R. Pott, good working conditions, technical infrastructure and a comprehensive library was provided.

I would like to thank PD Dr. G.-R. Walther for numerous valuable discussions and memorable field excursions during this study. PD Dr. G.-R. Walther initiated the study and supervised this thesis in cooperation with Prof. Dr. R. Pott.

Funding within the EU FP 6 Integrated Project "ALARM" (GOCE-CT-2003-506675) financially enabled the study. The "ALARM"-Project was also a platform for cooperation with numerous other scientists. Especially the cooperation with the working group of Prof. Dr. M.T. Sykes (Geobiosphere Science Centre, Department of Physical Geography & Ecosystems Analysis, Lund University, Sweden) was fruitful.

Parts of this study and related research which provided valuable knowledge for this study were funded by the German Research Foundation (DFG) (Project WA 1523/5-1) and the German Federal Agency of Nature Conservation (BfN) (FKZ 80581001).

Many thanks to the numerous persons who contributed with distribution records, climate data and other important information from all over the world, though most of them in Europe (see acknowledgements of the single chapters). Journal editors and peer-reviewers provided valuable advice that improved the single papers (see acknowledgements of the single chapters).

Thanks also to the "Doktorandengruppe" and Prof. Dr. H.P. Waldhoff for valuable support and enlightening discussions at different stages of the writing process. Finally, lots of thanks to colleagues and students at the Institute of Geobotany, especially Jana, Gunnar and Ann-Kathrin, and of course friends and family who encouraged me during the work on this study.

Last, but not least, thanks to Dennis, Brit and Bengt for all your patience and support.

Appendix

Appendix 1

Mean temperature values for the warmest and coldest month for a subset of climate stations used by Iversen (1944) and updated for the period 1971 – 2000.

	Climate station	surrogate station	warmest month		coldest month	
			Iversen (1944)	updated (1971-2000)	Iversen (1944)	updated (1971-2000)
Germany	Aachen		16.7	17.9	1.5	2.9
	Karlsruhe		19.0	20.1	0.4	1.9
	Kleve		17.0	17.7	1.0	2.6
	Gardelegen		17.6	17.7	-0.7	0.3
	Helmstedt		17.1	17.7	-0.7	0.6
	Wustrowa	Barth	16.8	16.8	-0.7	0.0
	Dessau	Wittenberg	17.9	18.5	-0.6	0.1
Denmark	Ringsted	Alstedgård II	16.5	16.7	-0.7	1.1
	Nykøbing	Abed	17.1	16.7	0.1	0.6
	Løgumkloster	FSN Skrydstrup	15.4	15,7	0.3	0,7
	Veljefjord	Billund Lufthavn	16.2	15.6	0.5	0.3
	Skanderborg	Ødum II	15.6	15.6	-0.3	0.3
	Hals	FSN Ålborg	16.1	16.0	-0.3	0.2
	Læsø	Skagen fyr	16.7	16,3	-0.4	1,0
Norway	Horten	Rygge	16.8	16.3	-2.4	-2.9
	Kristiansand	Oksøy fyr	16,2	15.6	-0.7	0.7
	Mandal	Mandal II	15.8	15.2	0.1	0.1
	Lindesnes		15.9	15.2	0.8	1.3
	Bergen	Bergen Florida	14.4	14.5	1.2	2.1
	Byrkenesøy	interpolated from Hellisøy and Takle	13.5	13.6	1.4	1.9
	Molde	Hustad II, extrapolated based on Tingvoll	13.9	12.9	0.2	1.1
	Kristiansund	Kristiansund, extrapolated based on Tingvoll	13.1	13.1	1.5	1.8
	Svolvær	Skrova fyr	12.7	12.7	-1.9	-0.3
Sweden	Vänersborg		16.4	16.1	-2.6	-1.7
	Borgholm	Ekologiska forskningsstationen	16.7	18.0	-1.2	-0.5
	Västeras		17.0	17.0	-3.8	-3.1
	Uppsala		16.5	17.3	-4.3	-3.1
	Visby		16.1	16.5	-0.9	-0.9

	warmest month	coldest month
Average deviation original stations	**0.4 ± 0.6**	**1.0 ± 0.5**
Average deviation surrogate stations	**-0.1 ± 0.5**	**0.7 ± 0.6**

Appendix 2

Detailed chronology of establishment of the new northernmost palm population (synonyms: *Trachycarpus fortunei* (HOOK.) WENDL. = *T. excelsa* WENDL. = *Chamaerops excelsa* THUNB.)

Time	Developmental stage
18th/19th century	In Europe, the hemp palm was first mentioned in 1795, and again in 1844 in England and Germany (JACOBI, 1998)

For southern Switzerland, the following records refer to *Trachycarpus fortunei*:

1882	Cultivated *Chamaerops excelsa* covered with plenty of **fruits** (KNY, 1882)
	(cf. Appendix S3 top left)
1904	Catalogues of spontaneously occurring and naturalised woody plants includes exotic species but no palms (BETTELINI, 1904, RIKLI, 1907, CHENEVARD, 1910, JÄGGLI, 1924)
1918	*Trachycarpus fortunei* survives the winter outdoor without any damage (BÄR, 1918)
1920	Subspontaneous occurrence of *T. fortunei* near Gandria (VOIGT, 1920)
1936	Blackbirds eagerly forage the fruits of *T. excelsa*. Seeds germinate easily, resulting in grasslike patches of **palm seedlings** below resting trees of birds (SCHRÖTER, 1936)
	(cf. Appendix S3 top right)
1950	*T. excelsa* locally colonises forests in Ticino on azonal sites such as shady and humid gorges (DAL VESCO, 1950)
1953	Description of lowland forests with few indigenous evergreen broad-leaved lower shrub species in an area along Lake Major, where at present, understorey vegetation is dominated by exotic evergreen broad-leaved species (KNAPP, 1953, cf. LÜDI, 1949)
1954/56	*T. excelsa* propagates in suitable habitats (humid gorges on southern-exposed slopes) comparable to indigenous species (SCHMID, 1956, cf. also LARCHER, 1954)
1961	Many of the cultivated species, such as e.g. *T. excelsa*, spread easily (SCHMID, 1961)
1964	Vegetation monographs on the lower forest belt mentioning escaped exotic evergreen broad-leaved species but no palms (OBERDORFER, 1964,

	ANTONIETTI, 1968)
1979	Occurrence of small palms* in the **herb layer** in zonal forests up to 480 m a.s.l. (ZUBER, 1979)

(cf. Appendix S3 bottom left)

1988	Occurrence of *T. fortunei* in the **shrub layer** (GIANONI et al., 1988)
1998	Strong expansion of evergreen broad-leaved species (incl. *T. fortunei*) is observed in the understorey vegetation of lowland forests (CARRARO et al., 1999)
2000	Fruiting individuals of *Trachycarpus*-palms on forest stands (WALTHER, 2000)
2003	**Establishment** of dense palm groves is a widespread phenomenon on southern exposed slopes up to approx. 400 m a.s.l.; occurrence of subspontaneous palm saplings at the max of ~ 800 m a.s.l. (WALTHER, 2003)

(cf. Appendix S3 bottom right)

* Although mentioned as *Chamaerops humilis* in the original literature, it could easily be mistaken for *Trachycarpus fortunei* in the juvenile stage as it was at that time (ZUBER, pers. comm.).

References

ANTONIETTI, A. (1968): Le associazioni forestali dell'orizzonte submontano del Cantone Ticino su substrati pedogenetici ricchi di carbonati. *Mitt. Schweiz. Anst. forstl. Vers.wes.* **44**: 85-226.

BÄR, J. (1904): Die Flora des Val Onsernone. *Beiträge zur geobotanischen Landesaufnahme der Schweiz* **5**.

BETTELINI, A. (1904): La Flora legnosa del Sottoceneri. – Tipografia e Litografia Cantonale, Bellinzona.

CARRARO, G., KLÖTZLI, F., WALTHER, G.-R., GIANONI, P. & R. MOSSI (1999): Observed changes in vegetation in relation to climate warming. Final Report NRP 31. – vdf, Hochschulverlag, Zürich.

CHENEVARD, P. (1910): Catalogue des plantes vasculaires du Tessin. – Kündig, Genève.

DAL VESCO, E. (ed.) (1950): Isole di Brissago – Parco botanica del cantone Ticino. – Arti grafiche Carminati, Locarno.

GIANONI, G., CARRARO, G. & F. KLÖTZLI (1988): Thermophile, an laurophyllen Pflanzenarten reiche Waldgesellschaften im hyperinsubrischen Seenbereich des Tessins. *Ber. Geobot. Inst. ETH, Stiftung Rübel, Zürich* **54**: 164-180.

JACOBI, K. (1998): Palmen für Haus und Garten. 4th ed. – BLV, München.

JÄGGLI, M. (1924): Cenni sulla Flora Ticinese. – Grassi & Co, Bellinzona.

KNAPP, R. (1953): Studien zur Vegetation und pflanzengeographischen Gliederung Nordwest-Italiens und der Südschweiz. *Kölner geogr. Arb.* **4**: 1-59.

KNY, L. (1882): Die Gärten des Lago Maggiore. – Parey, Berlin.

LARCHER, W. (1954) Die Kälteresistenz mediterraner Immergrüner und ihre Beeinflussbarkeit. *Planta* **44**: 607-638.

LÜDI, W. (1949): Bericht über den 6. Kurs in Alpenbotanik. *Ber. Geobot. Inst. ETH, Stiftung Rübel, Zürich* **12**: 12-50.

OBERDORFER, E. (1964): Der insubrische Vegetationskomplex, seine Struktur und Abgrenzung gegen die submediterrane Vegetation in Oberitalien und in der Südschweiz. *Beitr. naturk. Forsch. SW-Deutschl.* **23**: 141-187.

RIKLI, M. (1907): Zur Kenntnis der Pflanzenwelt des Kts. Tessin. *Berichte der Schweizerischen botanischen Gesellschaft* **17**: 27-63.

SCHMID, E. (1956): Flora des Südens. 2nd ed. – Rascher, Zürich.

SCHMID, E. (1961): Erläuterungen zur Vegetationskarte der Schweiz. *Beiträge zur geobotanischen Landesaufnahme der Schweiz* **39**.

SCHRÖTER, C. (1936): Flora des Südens. – Rascher, Zürich.

VOIGT, A. (1920): Beiträge zur Floristik des Tessins. *Berichte der Schweizerischen botanischen Gesellschaft* **26/29**: 332-357.

WALTHER, G.-R. (2000): Climatic forcing on the dispersal of exotic species. *Phytocoenologia* **30**: 409-430.

WALTHER, G.-R. (2003): Wird die Palme in der Schweiz heimisch? *Bot. Helv.* **113**: 159-180.

ZUBER, R.K. (1979): Untersuchungen über die Vegetation und die Wiederbewaldung einer Brandfläche bei Locarno (Kanton Tessin). *Beiheft zur Schweizerischen Zeitschrift für Forstwesen* **65**.

Appendix 3

Localities to the (non-exhaustive) compilation of *Trachycarpus*-sites based upon literature and internet search as well as personal observations and contacts (cf. Fig. 4.1 & 4.5).

Locality	Country	Source	Reference/Details
Naturalised			
Ticino	Switzerland	Walther 1999 & 2003	Botanica Helvetica 109, 153-167 & 113, 159-180
Austin, TX	USA	L. Lockett, pers. comm., 15.10.2005	
Wanganui	New Zealand	National Vegetation Survey (http://nvs.landcareresearch.co.nz/)	
South Auckland Land District	New Zealand	National Vegetation Survey (http://nvs.landcareresearch.co.nz/)	

6.7.1.1 Spreading outside gardens and parks

Bordeaux	France	P. Duhamel, written comm. http://www.philippeduhamel.fr/lepalmier_28_bordeaux.php see also: http://www.chez.com/palmiers/voir.php -> Région Bordelaise	Le Palmier No. 28
Alderney	UK	http://www.bsbiatlas.org.uk/main.php	
Lausanne	Switzerland	Bornand & Hoffer-Massard 2004; communicated by P. Vittoz, 10.10.2005	Bulletin du Cercle vaudois de Botanique 33, 99-122
Trsteno	Croatia	Kovacevic 1998	Acta Botanica Croatica 55/56, 29-40

Batumi	Georgia	Walther, pers. observ.	
Cheju	South Korea	Walther, pers. observ.	
Kanagawa Prefecture	Japan	Koike 2006 (http://vege1.kan.ynu.ac.jp/papers/4_4KoikeF.pdf)	
Victoria: Healesville, Creswic	Australia	Groves 1998	CRC for Weed Management Systems Technical Series No. 3
Nelson	New Zealand	Healy & Edgar 1980	Flora of New Zealand, Volume III, Hasselberg, Wellington

Self-seeding in gardens and parks

Zevergem	Belgium	van Speybroeck 2001	Chamaerops No. 42
Landevennec (Ile de Crozon)	France	M. Bourbier, written comm. (http://www.chez.com/palmiers/voir.php -> Bretagne)	
Pontivy	France	M. Bourbier, written comm. (http://www.chez.com/palmiers/voir.php -> Bretagne)	
Anduze	France	J.-M. Dupuyoo, pers. comm., 24.01.2006	
Heidelberg	Germany	F. Braun, written comm. (http://www.tiere-und-mehr.de/Pflanzen/Palmen/palmen.html)	Internetforum "Exoten und Garten"
Meran	Italy	L. Lutz, written comm., 06.04.2002 (http://exoten.dyndns.org/cgi-bin/archiv.cgi?function=2&index=4BD9914051F24DC8)	Reisebericht Meran/Südtirol, Teil 3
Riva	Italy	Berger & Walther 2006	Botanica Helvetica 116, 65-77
Montreux-Vevey	Switzerland	J.C. Wattenhofer, written comm. (http://www.fousdepalmiers.fr/temoignages.php)	

Zürich	Switzerland	D. Pfenninger, pers. comm., 27.08.2004 see also: Mackert & Hager 2004; Pfenninger 2004	Tages-Anzeiger 24. Juli 2004; 27. Juli 2004
Abbotsbury	UK	Walther, pers. observ.	
Torquay	UK	Walther, pers. observ.	
Trebah Garden	UK	Walther, pers. observ.	
Bournemouth	UK	C. Evans, pers. comm. 07.09.2004	
Anderson, SC	USA	M. Turner, pers. comm., 06.09.2005	
Victoria, BC	Canada	http://victoria.tc.ca/Environment/Exotic/palms/	The palms of Victoria
Nagano	Japan	Walther, pers. observ.	
Planted			
Vienna	Austria	M. Andresek, written comm. (http://land.heim.at/toskana/210025/palmstandort.html)	
Bad Deutsch Altenburg	Austria	Lackner 1995 & 1996	Chamaerops No. 19 & 24
Heist-op-den-Berg	Belgium	M. Vissers, written comm. (http://users.pandora.be/mvissers/)	
Kortrijk	Belgium	M. Vissers, written comm. (http://users.pandora.be/mvissers/)	
Musson	Belgium	J. Mazy, written comm., 05.03.2002 http://www.chez.com/palmiers/cold_/tra_for.php	Comptes rendus de la banque de Froid Section *Trachycarpus fortunei*
Tinlot	Belgium	F. Calmant, written comm., 11.03.2002 http://www.chez.com/palmiers/cold_/tra_for.php	Comptes rendus de la banque de Froid Section *Trachycarpus fortunei*

Location	Country	Reference
Plovdiv	Bulgaria	K. Donov, written comm. (http://www.polarpalm.net/)
Pula	Croatia	Tourism Office of Pula (http://www.pulainfo.hr/en/parkovi.asp)	Dendroflora of the public gardens of Pula in the past and today
Copenhagen	Denmark	http://davesgarden.com/pf/go/2255/	Dave's Garden; PlantFiles - Detailed information on Chusan Palm
Paris	France	van Speybroeck 2001	Chamaerops No. 42
Termes d'Armagnac	France	http://davesgarden.com/pf/go/2255/	Dave's Garden; PlantFiles - Detailed information on Chusan Palm
Lorient	France	M. Bourbier, written comm. (http://www.chez.com/palmiers/voir.php -> Bretagne)
Nice	France	http://www.chez.com/palmiers/voir.php -> Le midi
Quimper	France	M. Bourbier, written comm. (http://www.chez.com/palmiers/voir.php -> Bretagne)
Rayol Canadel	France	http://www.chez.com/palmiers/voir.php -> Le midi
Villefranche sur Saone	France	O. Blettery (http://www.chez.com/palmiers/voir.php -> Beaujolaise)
Isle de Batz	France	Prime 1992	Chamaerops No. 8
Betous, Gers	France	Ministère de la culture et de la communication 04/2004 http://www.culture.gouv.fr/culture/actualites/communiq/donnedieu/jardins2004.html
Marseille	France	Levy 1997	Chamaerops No. 25

Location	Country	Source
Saint Nazaire Les Eymes	France	R. de Looze, written comm., 15.04.2002 http://www.rarepalmseeds.com/pix/TraFor.shtml; see also http://www.palmeraiedesalpes.com/pres_gen.html
Cherbourg	France	http://www.ville-cherbourg.fr/fr/tourisme_decouverte/architecture_parc_et_jard/patrimoine_botanique/parcs_et_jardins/default.asp
Auxerre	France	P. Pauchard, written comm., 06.06.2002 http://www.chez.com/palmiers/cold_/tra_for.php Comptes rendus de la banque de Froid Section *Trachycarpus fortunei*
Brive-la-Gaillarde	France	M. Dosch, written comm., 03.05.2002 http://www.chez.com/palmiers/cold_/tra_for.php Comptes rendus de la banque de Froid Section *Trachycarpus fortunei*
Colmar	France	F. Perrichon, written comm., 07.10.2003 http://www.chez.com/palmiers/cold_/tra_for.php Comptes rendus de la banque de Froid Section *Trachycarpus fortunei*
la Celle	France	O. Martin, written comm., 17.04.2002 http://www.chez.com/palmiers/cold_/tra_for.php Comptes rendus de la banque de Froid Section *Trachycarpus fortunei*
Le Puy en Velay	France	L. Garrido, written comm., 22.03.2002 http://www.chez.com/palmiers/cold_/tra_for.php Comptes rendus de la banque de Froid Section *Trachycarpus fortunei*
Marennes	France	B. Binson, written comm., 24.06.2003 http://www.chez.com/palmiers/cold_/tra_for.php Comptes rendus de la banque de Froid Section *Trachycarpus fortunei*
Mont de Marsan	France	C. Faure, written comm., 22.03.2002 http://www.chez.com/palmiers/cold_/tra_for.php Comptes rendus de la banque de Froid Section *Trachycarpus fortunei*
Perigueux	France	M. Benouachkou, written comm., 15.03.2002 http://www.chez.com/palmiers/cold_/tra_for.php Comptes rendus de la banque de Froid Section *Trachycarpus fortunei*
Saussan	France	P. Vigroux, written comm., 09.04.2002 http://www.chez.com/palmiers/cold_/tra_for.php Comptes rendus de la banque de Froid Section *Trachycarpus fortunei*

Location	Country	Source	Notes
Loire - St Just St Rambert	France	B. Rassart, written comm., 15.04.2002	Comptes rendus de la banque de Froid
		http://www.chez.com/palmiers/cold_/tra_for.php	Section *Trachycarpus fortunei*
Villiers St Denis	France	L. Gillet, written comm., 03.03.2002	Comptes rendus de la banque de Froid
		http://www.chez.com/palmiers/cold_/tra_for.php	Section *Trachycarpus fortunei*
Vitry le Francois	France	F. Schneider, written comm., 10.08.2002	Comptes rendus de la banque de Froid
		http://www.chez.com/palmiers/cold_/tra_for.php	Section *Trachycarpus fortunei*
Hamburg W	Germany	http://www.palmen-exoten.de/articles_info.php?tPath=1_4&articles_id=5&language=de	
Leezen	Germany	http://www.palmen-exoten.de/articles_info.php?tPath=1_4&articles_id=5&language=de	
Unterhermsgrün	Germany	Höfer 2001	Chamaerops No. 43-44
Frankfurt	Germany	Gesell-Schulte 1995	Chamaerops No. 18
Langeoog	Germany	J. Barkowsky, pers. comm., 18.02.2005	
Düsseldorf	Germany	M. Lorek, written comm. (http://www.tropengarten.de/Botanik/duesseldorf.html)	
Karlsruhe	Germany	http://www.palmtrees.de/bilder/ka01.jpg	
Helgoland	Germany	J. Jäck, written comm., 24.06.2004 (http://velutina.brinkster.net/Helgopalmen/Helgoexoten.htm)	
Salzgitter	Germany	http://nixnutz.info/php/detail.php?detailspez=Trachycarpus&detailart=fortunei&detailwas=Palmen	
Simonswald	Germany	http://nixnutz.info/php/detail.php?detailspez=Trachycarpus&detailart=fortunei&detailwas=Palmen	
Tauberbischofsheim	Germany	http://www.thomasboller.de/	
Waren (Müritz)	Germany	http://nixnutz.info/php/detail.php?detailspez=Trachycarpus&detailart=fortunei&detailwas=Palmen	
Budapest	Hungary	A. Nemenyi, written comm., 22.12.1998	see also Chamaerops No. 29
		http://www.bg-map.com/palms/hungpalm.html	
Pecs	Hungary	A. Nemenyi, written comm., 22.12.1998 (http://www.bg-map.com/palms/hungpalm.html)	
Rosslare	Ireland	Fursman 1995	Chamaerops No. 20
Timoleague, County Cork	Ireland	Fursman 1995	Chamaerops No. 20

Location	Country	Reference	Source
South Tirol	Italy	Lackner 1996	Chamaerops No. 21
Rome Botanical Garden	Italy	Anonymus 2000	Chamaerops No. 43–44
Palermo	Italy	Plaumann 2001	Chamaerops No. 41
Florence	Italy	Karmelk 1994	Chamaerops No. 13
Luxembourg	Luxembourg	http://webplaza.pt.lu/lucioiob/trachycarpus.htm; see also: http://www.exoticsinluxembourg.com	
Sör-Hidle	Norway	S. & O. Bryn (http://www.florogfjære.no/palmer.html)	Flor & Fjære
Lisbon	Portugal	Palms and Gardens in the South-west of Europe (http://www.marblenet.es/pjse/listapeng.htm)	
Salema (Algarve)	Portugal	Wijchgel 1998	Chamaerops No. 31
Bilbao	Spain	Walther, pers. observ.	
Elche	Spain	Kyburz 1995	Principes 39(4)
La Palma	Spain	Croft 2002	Chamaerops No. 45
Teneriffe	Spain	Morici 2001	Chamaerops No. 42
Skanör	Sweden	L. Klingstrom, written comm. http://snowpalm.dyndns.org/eng/palm.html	Palms for Scandinavia
Gothenburg	Sweden	Drakenius 1997	Chamaerops No. 29
Ängelholm	Sweden	L. Klingstrom, written comm. 25.12.1999 (http://www.snowpalm.dyndns.org/eng/)	
Basel	Switzerland	http://www.winterhart.ch/suche.php -> Trachycarpus; see also: http://pages.unibas.ch/botgarten/mittelmeergarten/b.shtml	
Fläsch	Switzerland	Walther, pers. observ.; see also: http://www.winterhart.ch/suche.php -> Trachycarpus	
Gröne	Switzerland	H. Casal, written comm., 22.01.2004 http://www.chez.com/palmiers/cold_/tra_for.php	Comptes rendus de la banque de Froid Section *Trachycarpus fortunei*
Spiez	Switzerland	http://www.winterhart.ch/suche.php -> Trachycarpus	

Location	Country	Reference
Scilly islands	UK	J. Ash, pers. comm., 18.10.2004 Email correspondence
Colchester	UK	http://davesgarden.com/pf/go/2255/ Dave's Garden; PlantFiles - Detailed information on Chusan Palm
Oxford	UK	JungleSeeds&Gardens (http://www.junglegardens.co.uk/PlantOrders/enter.html)
Edinburgh	UK	Royal Botanic Garden Edinburgh (http://www.rbge.org.uk/rbge/web/news/season.jsp)
St. Andrews Botanical Garden	UK	St Andrews Botanic Garden (http://www.st-andrews-botanic.org/) Plant of the Month July 2002
Dover	UK	Brown 1994 Chamaerops No. 14
Belsay Hall	UK	English Heritage; Belsay Hall – Northumberland (http://www.english-heritage.org.uk/server/show/nav.0010020080000800b)
Worthing	UK	B.M. Lamb, written comm.
Newport	UK	Kealaher 1992 Chamaerops No. 7
Cheshire	UK	Oaks 2001 Chamaerops No. 42
Jersey	UK	Carter 2000 Chamaerops No. 40
London	UK	Dixon 1995 Chamaerops No. 17
Swansea	UK	Dixon 1995 Chamaerops No. 17
Barnsley	UK	Darlow 1994 Chamaerops No. 13
Lleyn	UK	Brown 2000 Chamaerops No. 37
Kintyre	UK	Ferguson 1998 The Kintyre Mag 22, 5 http://www.kintyremag.co.uk/1998/22/page5.html
Inverewe, Poolewe	UK	Payne 1992 Chamaerops No. 8
Mt Stewart, Strangford	UK	G. & F. McCormick, written comm. (http://www.gmcc0266.plus.com/britpalm.htm)

Sukhumi	Abkhazia/ Georgia	communicated by Guy through Northwest Palms Discussion Board, 26.08.2005	
Sochi	Russia	H. Rossel, written comm. (http://www.hansrossel.com/fotos/fotografie/rusland/rus_d3150.htm)	
Crimean Coast	Ukraine	T.W. Cheng, written comm. 04.09.1999 (http://weecheng.com/europe/bbs/crimea/livadia2.jpg)	
Caspian Sea coast between Baku and the Iranian border	Azerbayjan	communicated by Guy through Northwest Palms Discussion Board, 26.08.2005	
Ashgabat	Turkmenistan	communicated by Guy through Northwest Palms Discussion Board, 26.08.2005	
Brandmuellers Farm, Johannesburg	South Africa	K. Peper, written comm. (http://www.arboretum.co.za/)	List of Plants
Rhode Island	USA	http://members.cox.net/joeb9/	Growing hardy palms and subtropical in Rhode Island
Southeastern Pennsylvania	USA	D. Levan, written comm., 01.08.2005 http://www.bg-map.com/palms/SE_PA.html	Windmill palm in southeastern Pennsylvania
Atlanta, GA	USA	http://davesgarden.com/pf/go/2255/	Dave's Garden; PlantFiles - Detailed information on Chusan Palm
Canoga Park, L.A., CA	USA	http://davesgarden.com/pf/go/2255/	Dave's Garden; PlantFiles - Detailed information on Chusan Palm
Clifton, CO	USA	http://davesgarden.com/pf/go/2255/	Dave's Garden; PlantFiles - Detailed information on Chusan Palm
Edmond, OK	USA	http://davesgarden.com/pf/go/2255/	Dave's Garden; PlantFiles - Detailed information on Chusan Palm

Fayetteville, NC	USA	http://davesgarden.com/pf/go/2255/	Dave's Garden; PlantFiles - Detailed information on Chusan Palm
Juneau, AK	USA	http://davesgarden.com/pf/go/2255/	Dave's Garden; PlantFiles - Detailed information on Chusan Palm
Lake Worth, FL	USA	http://davesgarden.com/pf/go/2255/	Dave's Garden; PlantFiles - Detailed information on Chusan Palm
Mont Vernon, NH	USA	http://davesgarden.com/pf/go/2255/	Dave's Garden; PlantFiles - Detailed information on Chusan Palm
Murphysboro, IL	USA	http://davesgarden.com/pf/go/2255/	Dave's Garden; PlantFiles - Detailed information on Chusan Palm
Natchez, MS	USA	http://davesgarden.com/pf/go/2255/	Dave's Garden; PlantFiles - Detailed information on Chusan Palm
Oklahoma City, OK	USA	http://davesgarden.com/pf/go/2255/	Dave's Garden; PlantFiles - Detailed information on Chusan Palm
Portland, OR	USA	http://davesgarden.com/pf/go/2255/	Dave's Garden; PlantFiles - Detailed information on Chusan Palm
Rio Rancho, NM	USA	http://davesgarden.com/pf/go/2255/	Dave's Garden; PlantFiles - Detailed information on Chusan Palm
San Mateo, CA	USA	http://davesgarden.com/pf/go/2255/	Dave's Garden; PlantFiles - Detailed information on Chusan Palm
Sparks, NV	USA	http://davesgarden.com/pf/go/2255/	Dave's Garden; PlantFiles - Detailed information on Chusan Palm

Thousand Oaks, CA	USA	http://davesgarden.com/pf/go/2255/	Dave's Garden; PlantFiles - Detailed information on Chusan Palm
Wichita, KS	USA	http://davesgarden.com/pf/go/2255/	Dave's Garden; PlantFiles - Detailed information on Chusan Palm
Woodbury, CT	USA	http://davesgarden.com/pf/go/2255/	Dave's Garden; PlantFiles - Detailed information on Chusan Palm
Birmingham, AL	USA	P.T. Nawrocki, written comm. 01/2001 http://www.thewisegardener.com/Articles/Palm_0f_The_Month/Trachycarpus_fortunei.htm	January, 2001's Palm of the Month: Trachycarpus fortunei
Columbia, SC	USA	P.T. Nawrocki, written comm. 01/2001 http://www.thewisegardener.com/Articles/Palm_0f_The_Month/Trachycarpus_fortunei.htm	January, 2001's Palm of the Month: Trachycarpus fortunei
Bethany Beach, DE	USA	Central Florida Palm & Cycad Society http://www.plantapalm.com/centralfl/ObservationSpecies.asp?PalmID=458	CFPACS - Observations for Trachycarpus fortunei
Lake Placid, FL	USA	Central Florida Palm & Cycad Society http://www.plantapalm.com/centralfl/ObservationSpecies.asp?PalmID=459	CFPACS - Observations for Trachycarpus fortunei
Longwood, FL	USA	Central Florida Palm & Cycad Society http://www.plantapalm.com/centralfl/ObservationSpecies.asp?PalmID=460	CFPACS - Observations for Trachycarpus fortunei
Clarkston, WA	USA	I. Barclay, 1998 (http://www.plantapalm.com/vpe/photos/Species/trachycarpus_fortunei.htm)	
Waterford Township, South Jersey	USA	Southern Accents 1999 (http://members.aol.com/fdrapa/windmill.htm)	
Poteau	USA	Oklahoma Gardening 2003 http://home.okstate.edu/Okstate/dasnr/hort/hortlahome.nsf/toc/november	Oklahoma Gardening Information Sheet (#3020)

Dallas, TX	USA	Cerbone 2002	Chamaerops No. 45
Coos Bay, OR	USA	I. Barclay, written comm. (http://www.angelfire.com/bc/eucalyptus/iansplants/gallery/46/)
Kentucky	USA	http://palmsnorth.com/forum/viewtopic.php?p=666&sid=cef4a72d8baf266d07ea3605ebc33cfc
Vancouver	Canada	http://davesgarden.com/pf/go/2255/	Dave's Garden; PlantFiles - Detailed information on Chusan Palm
Xalapa	Mexico	Andersson 1999	Chamaerops No. 34
Porto Alegre	Brasil	Walther, pers. observ.	
Perth	Australia	Shelton 1993	Chamaerops No. 9
Tasmania	Australia	P.T. Nawrocki, written comm. 01/2001	January, 2001's Palm of the Month: Trachycarpus fortunei http://www.thewisegardener.com/Articles/Palm_0f_The_Month/Trachycarpus_fortunei.htm
Armidale, NSW	Australia	Oakes 2005	Chamaerops No. 50
Canterbury Land District	New Zealand	New Zealand Plant Database (http://nzflora.landcareresearch.co.nz)	
Invercargill	New Zealand	Jackson 1993	Chamaerops No. 12

Appendix 4

Pictures illustrating the different developmental stages of the new northernmost *Trachycarpus* population.

Trachycarpus fortunei (HOOK.) WENDL.

top left: infructescence; *top right*: palm seedlings;
bottom left: subspontaneous palm sapling; *bottom right*: dense palm stands

Appendix 5

Realised and modelled (grey shaded areas) distribution of *Trachycarpus fortunei* in its native range, using the following parameters: 2.2°C as the lower limit of the monthly mean temperature of the coldest month (Tcold) and GDD_5 = 2300, a maximum mean temperature of the coldest month of 15.5°C and a tolerated drought index of 0.26 (for details see text and Fig. 4.5).

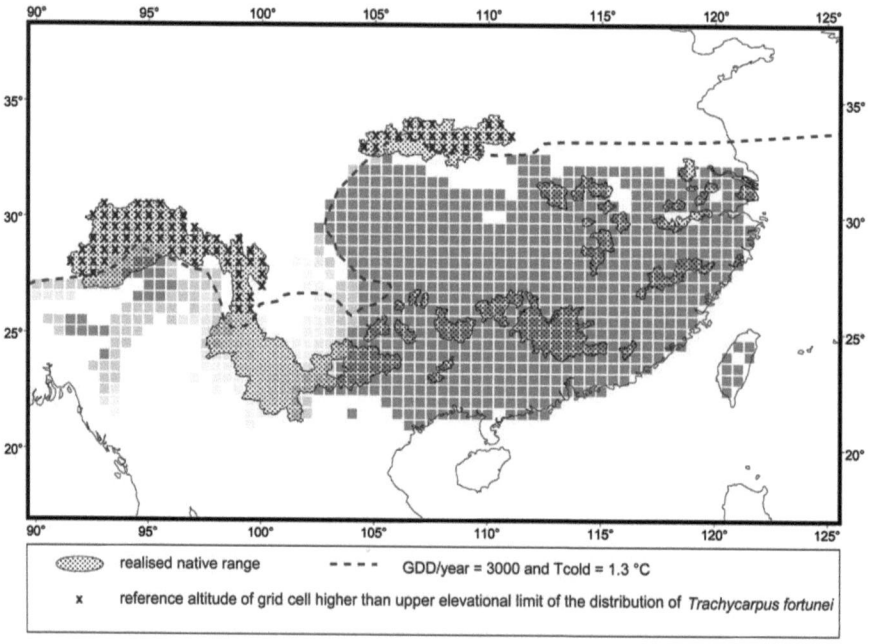

Die VDM Verlagsservicegesellschaft sucht für wissenschaftliche Verlage abgeschlossene und herausragende

Dissertationen, Habilitationen, Diplomarbeiten, Master Theses, Magisterarbeiten usw.

für die kostenlose Publikation als Fachbuch.

Sie verfügen über eine Arbeit, die hohen inhaltlichen und formalen Ansprüchen genügt, und haben Interesse an einer honorarvergüteten Publikation?

Dann senden Sie bitte erste Informationen über sich und Ihre Arbeit per Email an *info@vdm-vsg.de*.

Sie erhalten kurzfristig unser Feedback!

VDM Verlagsservicegesellschaft mbH
Dudweiler Landstr. 99
D - 66123 Saarbrücken

Telefon +49 681 3720 174
Fax   +49 681 3720 1749

www.vdm-vsg.de

Die VDM Verlagsservicegesellschaft mbH vertritt

Printed by Books on Demand GmbH, Norderstedt / Germany